Sounds

Sounds

A Philosophical Theory

Casey O'Callaghan

OXFORD
UNIVERSITY PRESS

OXFORD
UNIVERSITY PRESS

Great Clarendon Street, Oxford OX2 6DP

Oxford University Press is a department of the University of Oxford.
It furthers the University's objective of excellence in research, scholarship,
and education by publishing worldwide in

Oxford New York

Auckland Cape Town Dar es Salaam Hong Kong Karachi
Kuala Lumpur Madrid Melbourne Mexico City Nairobi
New Delhi Shanghai Taipei Toronto

With offices in

Argentina Austria Brazil Chile Czech Republic France Greece
Guatemala Hungary Italy Japan Poland Portugal Singapore
South Korea Switzerland Thailand Turkey Ukraine Vietnam

Oxford is a registered trademark of Oxford University Press
in the UK and in certain other countries

Published in the United States
by Oxford University Press Inc., New York

First published 2007

British Library Cataloguing in Publication Data

Data available

Library of Congress Cataloging in Publication Data

Data available

Typeset by Laserwords Private Limited, Chennai, India
Printed in Great Britain
on acid-free paper by
Biddles Ltd, King's Lynn, Norfolk

ISBN 978–0–19–921592–8

10 9 8 7 6 5 4 3 2 1

Preface

My work on sounds begins with an interest in perception. I want to know, in particular, how perceptual experience shapes our understanding of the natures of what we perceive. I have always thought that the problems of color encapsulate what is most captivating about philosophy. It strikes me, however, that the vast majority of philosophical attention to perception has focused on vision, and that thinking predominantly about the visual drives so much philosophical theorizing about what we perceive. Though this is unsurprising given that vision is so central to our conception of the world, and of our place in it, such narrow attention is bad policy in developing a comprehensive, general account of perception. I am skeptical that what we say about color and vision extends neatly to sounds and the other senses.

A broadened understanding of perceptual modalities opens new doors in developing the theory of perception. It draws attention to unrecognized features of perceptual experience and provides persuasive lessons against otherwise favored accounts. It compels us to acknowledge differences in the roles of the perceptual modalities and the significance of the relationships among them. I believe, as I claim in this book, that it thereby reveals what is most crucial to perceiving.

What I have discovered in sounds and auditory perception is a nearly untouched terrain rich in philosophical problems and illumination, with a growing body of empirical research to inform philosophical thought. This book is my start at exploring the philosophy of sounds and auditory experience. My goal is to present problems unique to the philosophy of sounds, and to draw attention to those places where audition confirms or conflicts with what we think about other perceptual modalities and their objects. Because vision provides my primary contrast case, I move only incrementally beyond the focus that motivates this work.

My main target here is the nature and perception of ordinary environmental sounds, such as bangs, rattles, beeps, clicks, chirps, and rustles. It might surprise some that I do not address, at length, the phenomenon of speech perception. Speech is a complex, vexed topic that deserves its own book-length philosophical treatment. Furthermore, it is not immediately evident either how speech sounds bear upon the metaphysics of ordinary sounds or how perceiving speech impacts our account of the perception of ordinary environmental sounds. Speech perception, and, in particular, the seemingly perceptual grasp of words, their force, and their content, likely involves perceptual mechanisms that are, in significant measure, distinct from those that deal with other sorts of environmental sounds. Speech perception is not, after all, afforded simply by the capacity to hear in one's environment—many animals perceive sounds but do not perceive speech. So I have kept from attempting a full account of the sounds of speech, and of perceiving speech. Nonetheless, the theory I propose, according to which sounds are events, and my conclusions concerning cross-modal illusions have natural applications to the philosophical problems of speech.

Similarly, I do not attempt to develop an account of music or musical experience. I leave that, for now, to the philosophers and cognitive scientists of music. I hope that my remarks about the nature of sounds and auditory perception will enrich their discussion.

The landscape this small move reveals leaves work enough. This book is not my attempt to reconcile auditory perception with a general theory of perception. Though certain themes will become evident, and I make proposals concerning how to move ahead, I hope for now to provide a framework for that dialogue. My aim here is to bring into view those features of sounds and auditory experience that inform metaphysical and perceptual theorizing, and that constitute the central problems of the philosophy of sound.

C.O.

Acknowledgements

I am grateful to quite a number of teachers, colleagues, and correspondents, and I owe them special thanks. A number of individuals have been particularly generous with conversation, commentary, support, and ideas. I express my gratitude to Paul Benacerraf, Sarah Broadie, John Burgess, Roberto Casati, David Chalmers, David Cummiskey, John Doris, Gilbert Harman, Benj Hellie, Mark Johnston, the late David Lewis, Brian Loar, Brian McLaughlin, Peter Momtchiloff, Matthew Nudds, Mark Okrent, Robert Pasnau, Roger Scruton, Susanna Siegel, Jeffrey Speaks, John Strong, Matthew Stuart, Scott Sturgeon, and Thomas Tracy. I expressly thank Alex Byrne, Sean Kelly, Alva Noë, and Gideon Rosen for extensive written comments and for conversations that have affected my views about both sounds and perception. Each has influenced the timbre of this work.

Audiences to whom I presented material from this book offered very helpful questions and challenges, as well as instructive opportunities to develop my ideas about sounds and auditory perception. This work owes much of its present shape to them. If worries and objections remain unsatisfactorily addressed, it is not because these astute audiences failed to raise them. I thank audience members at Princeton University, Florida State University, University of California—Santa Barbara, University of St. Andrews, N. E. H. Institute on Consciousness and Intentionality, University of California—Santa Cruz, Auburn University, University of London, University of Manitoba, Dartmouth College, Bowdoin College, University of Georgia, Toward a Science of Consciousness, Society for Philosophy and Psychology, Boston Colloquium for Philosophy of Science, University of Maine, Orono, University of Notre Dame, Massachusetts Institute of Technology, and Washington University in Saint Louis. Students in my seminar on colors and sounds during fall 2004 and 2006 also offered sharp remarks on portions of the manuscript.

Matthew Côté expertly composed the figures in Chapter 6. Sachi Cole and Claire Guyton provided valuable editorial and bibliographic assistance.

Portions of Chapter 9 are reworked from my paper 'Echoes', which appears in *The Monist* 90: 3 (July 2007). Portions of Chapter 3 draw on 'Perceiving the Locations of Sounds', which appears in the *European Review of Philosophy* 7 (2007). I gratefully acknowledge permission to incorporate this material.

Bates College granted research funding that supported the writing of this book.

Friends and family have been tremendously supportive, witty, and fun. Thanks to Antony Eagle, Maree Henwood, Jackson, Scott Jenkins, Daniel and Stefanie Kelly, Francis Kelly, Margaret Kelly, Michael and Allana Kelly, Laura Koehn, Linda and Martin Koehn, Melissa Koehn, Marlene Lynch Ford, Vanessa Nesvig, Anthony and Pamela O'Callaghan, Elyse Speaks, and the late Frances Spaulding.

Simon Keller has been a loyal friend and philosophical confidant throughout.

I cannot imagine having written this book without Emily Koehn's immeasurable love and friendship. She spins songs from sounds, and remains the most devout event theorist.

I dedicate this book to my grandfather, John Avery Spaulding.

Contents

List of Figures

By listening, one will learn truths.
By hearing, one will only learn half truths.
Lucky Numbers 6, 14, 19, 27, 30, 34.

From a fortune cookie

A sound does not view itself as thought, as ought, as needing another sound for its elucidation, as etc.; it has not time for any consideration—it is occupied with the performance of its characteristics: before it has died away it must have made perfectly exact its frequency, its loudness, its length, its overtone structure, the precise morphology of these and of itself.

John Cage

1

Sonic Realism

1.1 The Tyranny of the Visual

Imagine standing in a bowling alley. Facing the pins, poised for the delivery, consider what you perceive. You might see rows of glossy lanes and racks of oblong pins, bowlers sliding toward foul lines in your periphery, your hands around a translucent ball, and, perhaps, the unusual shoes on your feet. That, of course, is not all you see; and it is not all you perceive. You feel the heft of the ball in your arms and its cool touch on your palms and fingers. You smell aerosol sanitizing spray in the air, and traces of the odors it fails to smother. You hear voices and banter. And you hear the expectant, droning sound of journeys up lanes, which occasionally culminate in a resounding, clanging pileup. If it is afternoon, perhaps you hear only ventilation and the sound of your own breathing.

Perception has drawn philosophical attention since antiquity. Plato distinguished appearances from reality—and inflated this distinction's consequences concerning what exists and what can be known through the senses. Descartes spurred centuries of debate over whether experience reveals things beyond the mind by conceiving a wholesale illusion spawned by an evil demon. Gilbert Harman, at the end of the last century, touched off a new storm of discussion about what is revealed in sensory awareness with the following remarks about the *transparency* of perceptual experience.

When you see a tree, you do not experience any features as intrinsic features of your experience. Look at a tree and try to turn your attention to intrinsic features of your visual experience. I predict you will find that the only features there to turn your attention to will be features of the presented tree. (Harman 1990: 39)

Perception engages philosophers because it purports to form a subject's primary mode of access to the world—it furnishes the materials of experience, grounds thought, and guides action. Perception shapes our understanding of things and events in the world and provides the data according to which experience can be evaluated as accurate, illusory, or misleading. Contemporary advances in natural and neural sciences only reinforce the importance of understanding the relationship between the world as it is characterized and explained through scientific inquiry and the world as it is encountered and experienced by perceiving creatures. The stories frequently differ so dramatically. Empirically informed philosophical theorizing is essential to investigating the relationships between the manifest and scientific images of the world.

Philosophical thinking about perception has been shaped to a remarkable extent by attention to vision. Humans, we frequently hear, are visual creatures, and vision has not disappointed philosophers as a source of insight into perception. At the outset of philosophy's modern era, Kepler (1604/2000), Descartes (1637/2001), Newton (1704/1979), and Berkeley (1709/1975*a*) each developed theories of optics and vision. The contemporary history of research on vision has fueled intense philosophical activity concerning the perception of colors, objects, motion, and causation.[1] 'Visuocentrism' has shaped our understanding of perception and its role. Vision has furnished the phenomena, questions, and puzzles with which philosophical theorizing about perception must deal. Thought experiments dealing with color spectrum inversion and Mary the blind color scientist, as well as phenomenological and empirical facts involving the waterfall illusion, blindsight, and

[1] Landmark empirical works on these topics, which have influenced philosophical thinking, include Helmholtz (1925), Michotte (1963), Land (1977), Gibson (1979), Adelson and Movshon (1982), Marr (1982), Rock (1983), and S. Palmer (1999).

change and inattentional blindness, all have driven philosophical views about perception in recent years.[2] The philosophical literature dealing with visual perception and consciousness is staggeringly huge. Even the terminology used to conduct philosophical debate about perceptual experience—*appearance, scene, image, observe*—is predominantly visual. Neutral or non-visual language—*recognize, discern, perceive*—has acquired a palpably visual tinge.

Hearing, however, also is a rich source of perceptual information about one's environment. Currently I hear the sound of a herring gull squawking outside a window I am facing. I hear a truck passing from left to right on the street below and behind me. I have heard the sound of my neighbor vacuuming through the wall during the past few minutes. Just by way of hearing, you are able to discern what kinds of things are around, what is happening to those things, how long these activities last, and where all of this occurs. Recall the bowling alley. You hear when a ball has gone in a gutter, how long it rolls, and when it has made contact with the pins.

I do not believe that the visuocentric focus has been entirely misguided. Instead, I think it is clear that this focus is too narrow if we hope to make further progress in thinking about perception. I want, therefore, to propose that hearing and the world of sounds are rich with raw material that presents both novel philosophical problems and telling new instances of old ones. The case of sounds and audition demonstrates that attention to modalities other than vision enriches our understanding of perception and its role in situating oneself, forming beliefs, and acting upon the environment.

This perspective contrasts sharply with a line of thought implicit in the history of philosophical work on perception. The more or less implicit assumption has been that what we learn about perception by studying vision generalizes to the other sense modalities. Put another way, vision is the representative paradigm of perception

[2] Shoemaker (1982) and Block (1990); Jackson (1986); Crane (1988); Weiskrantz (1986); and Noë (2002) are respective starting points for the discussions I mention here.

and holds the key to understanding the nature and purpose of perceiving. According to the traditional line of thought prominent from the early modern era to the present, in the philosophically interesting respects at least, as things are with vision, so they are with hearing, touch, and olfaction. The perceptual modalities have been treated as analogous in that, from the perspective of a philosophical account of perception, understanding auditory, tactile, or olfactory perception involves little more than extrapolating or transposing from an account of vision. A line of thought not accidentally related has been particularly strong in the case of the secondary or sensible qualities. The assumption is that as things are with colors, so they are with sounds, tastes, and smells.

This book is predicated on skepticism about this kind of claim. Not only does each of the various perceptual modalities and its objects warrant philosophical interest in its own right, but attention to non-visual modalities may force reexamination of visuocentric hypotheses about the nature and character of perception. It may compel us to alter our account of the epistemology of perception or to shift our stance on perception's role in mediating thought and action. More modestly, such attention makes good philosophical methodology, given what we already have learned from thinking about vision. What I urge is that we put to rest the traditional lines of thought and unleash the philosophies of the senses and their objects. In short, that we end the tyranny of the visual.

It is in this spirit that I develop in this book a philosophical account of sounds and their perception.

1.2 Sounds and Visuocentrism

Visual experience most notably involves awareness of commonplace objects, items, and stuffs. Sighted subjects seem to see chairs, trees, coffee, cars, and persons during ordinary day-to-day visual experience. My current visual experience consists in awareness of walls, doors, desk, cup, computer, lamp, hands, and so on, in ever-increasing detail. Though we often seem to see things that are not ordinary objects, such as the sky, lightning, shadows, and glare, vision presents them as if they had more in common with typical material objects than they do. Though we see events, there is an important sense in which it seems to one that seeing an event depends upon seeing its participating objects. Seeing the chair's collapse involves seeing the chair, its parts, and the floor.

Visual experience, furthermore, presents objects and quantities of stuff as having or possessing such attributes as colors, shapes, and sizes. The visible features that characterize visible objects and stuffs are qualities that seem bound up with those objects. The visible red color of a stop sign seems 'stuck on' the surface of the stop sign. Its octagonal shape is not separable from the sign itself in visible experience. Part of vision's central task is to furnish awareness of things and their attributes.

Taking vision to exemplify perception's function, however, may lead to puzzlement over the nature and place of sounds in the world. Sounds themselves are not good examples of ordinary objects or substances. Sounds do not seem in audition to have detailed shapes and sizes; they do not seem to have mass, to be solid, or to flow. One cannot grasp or sit upon a sound.

If perceiving serves to acquaint us with the world of material things and their properties, then it is natural to think that sounds must be audible attributes of the same objects we see. Philosophical tradition in fact regards sounds in this way. Sounds are counted with colors, tastes, and smells among the secondary or sensible attributes of objects, and inherit the problems of the more frequently discussed sensible qualities.

But though we speak of the sound of a glass breaking or the sound of a tuba, it is not clear that sounds seem bound to their purported bearers in the manner of colors, shapes, textures, and tastes. And this is not just a matter of the more fleeting natures of sounds. The immediacy of one's auditory awareness of ordinary objects does not match that of vision, and there is the sense that in hearing one is at an additional remove from everyday objects and the events in which they participate. This sense is apparent in implicit standards of evidence. Seeing is believing, we say, but don't believe everything you hear. Seeing Elvis in the flesh holds evidential weight we do not grant to merely hearing him in the next room.

Sounds as we experience them in hearing are audibly independent from ordinary material objects in a way that colors and shapes are not visibly independent from objects. The sounds seem produced or generated by ordinary objects and events; colors and shapes do not. Apart from bearing on matters of evidence, this fact makes mysterious the place of sounds in the world for a visuocentric understanding of perception. If perception is for revealing objects and their properties, and sounds seem among neither objects nor their properties, it is tempting to suppose that sounds are mysterious, ethereal, or otherwise questionable items of sense. If sounds as we perceive them do not exhibit the common marks of items in the material world, and if they are not obviously features of those items, that might encourage us to believe that sounds have no natural home in the world. That, in turn, may tempt us to understand sounds as having no place other than the mind.

Similar thinking perhaps encouraged early modern philosophers to count the sounds with the colors as sensory qualities or purely mental items. Evidence of the attitude still abounds. In the most famous philosophical discussion of sounds, P. F. Strawson (1959) argues that objects are fundamental to our conceptual scheme and claims that a world of sounds would be a non-spatial world. Combined with a version of the Kantian thesis that experience of an objective world requires space, the world of sounds—in contrast to the world of vision—leaves no room for non-solipsistic consciousness. Discussing Strawson's argument, Gareth Evans claims, 'the truth of a proposition to the effect that there is a sound at such-and-such a position must consist in this: if someone was to go to that position, he would have certain auditory experiences' (Evans 1980: 274). Propositions about sounds on this view reduce to propositions about experiences.

D. L. C. Maclachlan arrived at the view that sounds are mental by comparing hearing to the experience of pain.

Like the pain I feel when the mosquito bites, the noise I hear when the dog barks is also an effect produced in me by the creature concerned. The sounds I actually hear are as private as the pains I feel. There are therefore good reasons for grouping together the experienced sounds and the experienced pains and calling them sensations. (Maclachlan 1989: 30)

Sounds, according to this account, are sensations caused by things and events in the world, but do not inhabit that world. Just as the pain of the bite does not belong to the mosquito, neither does the buzzing sound belong to the bug. It, like the pain, belongs to the bitten.

R. M. P. Malpas struggles to explicate the expression 'the location of sound' as follows:

I do not mean by 'location' 'locality', but 'the act of locating', and by 'the act of locating' I do not mean 'the act of establishing in a place', but 'the act of discovering the place of'. Even so 'location' is misleading, because it implies that there is such a thing as discovering the place of sounds. Since sounds do not have places there is no such act. (Malpas 1965: 131)

Philosophical axes and analyses aside, Matthew Nudds nicely sums up puzzlement about the place of sounds in the world by quoting Jonathan Rée.

In a recent book Jonathan Rée tells of how as a child he used to wonder which would be worse: to lose one's sight or one's hearing? Much worse, he concluded, to lose one's sight, since

> Sounds seemed to me to be nature's waifs and strays: they did not fit into the familiar world of physical things, and they could not be tracked down by my other senses either ... They were not part of the material world, and they had no weight to them, no substance. Is it surprising that I thought I could happily do without them? (Rée 1999: 19)

Whether or not we would be happy to do without sounds, the idea that our experience of sounds is of things which are distinct from the world of material objects can seem compelling. All you have to do to confirm it is to close your eyes and reflect on the character of your auditory experience. (Nudds 2001: 210)

If vision is the paradigm for perception, and perceptual awareness involves in the first instance the experience of ordinary material things and their attributes, then it is no surprise that sounds should be suspect (and favored terrain of enterprising idealists). And it is no wonder that some have understood the world of sounds as a world without space. For sounds do not seem in auditory experience like ordinary objects, and we do not hear them to qualify ordinary objects in the way that visible attributes do. The world of sounds confounds visuocentric thought.

1.3 Toward Sonic Realism

Vision, in fact, seems a dubious paradigm if we hope to understand all the senses as ways of encountering, experiencing, and perceiving one's environment. At first impression, it is shocking that considerations relevant to theorizing philosophically about smells, odors, pitches, or warmth should mirror those that impact the theory of colors, brightness, or visible shapes. What is striking is the diversity of experience, and that experience reveals so much variety. Awareness in heterogeneous modalities discloses items that appear as particulars or as attributes, though not necessarily as ordinary objects or their attributes. Such items and features may appear bound to particulars, but they may seem produced or generated by distinct entities. They may seem located in three-dimensional space or to have no sensible location. The information our senses furnish about commonplace material objects and happenings may seem more direct or at great remove. Their qualities may constitute spaces with one dimension or with many, and they exhibit seemingly incommensurable characteristics. If we hope to discover what it is to perceive or to discern the natures of the things we sense, devoting attention to the umpteen ways of experiencing is unavoidable. A clearheaded philosophy respects where experiences and their objects are alike and recognizes where they differ.

Abandoning visuocentrism frees us to regard sounds as of interest not only for what they reveal about the world and about what is visible, but also as entities in their own right. Freedom from focus on the visual encourages us to eliminate confusion about the place of sounds and to capture the character and usefulness of auditory awareness. This does not mean we should be indifferent to what unifies perceptual experience, or to what constitutes perception's nature and distinctive function. Rather, we should resist uncritically permitting a privileged modality to drive the unifying explanations where we risk leaving out important facets of what we aim to explain.

If sounds are mysterious, ethereal, or mental artifacts, it is in fact difficult to grasp why it might be so useful to perceive them. Taking sounds seriously does not mean considering them in isolation. Audition furnishes 360° awareness of a three-dimensional spatial field and directs the orientation of visual attention toward the sources of perceived sounds. Awareness of sounds provides information about the sorts of objects that populate the environment, about what those objects are doing, and about how they interact. Audition surpasses vision in the ability to detect change and in the ability to monitor multiple sources of information. These capacities suit audition for the perception of speech, a task that for most of us frustrates unaided vision. Investigating the nature of sounds, as we perceive them, without excessive pressure from the visual model allows us to make sense of how hearing functions and why it functions as it does.

Suppose sounds are not merely mental artifacts of sensation. Suppose that a sound can seem distinct from oneself, and that subjects sometimes really do hear sounds. Sounds, that is, frequently are the targets of a perspective. Realism about sounds—*sonic realism*—is the view that the world contains sounds whose existence is not entirely dependent upon the auditory experiences of subjects. Realism about sounds as I shall develop it maintains, furthermore, that the world of sounds is the same world that contains ordinary material objects, events, and their attributes. The world of sounds is not a distant realm occupied by alien entities or attributes. Sounds are in the world. Sounds are the entities that, in the first instance, we auditorily perceive. Since perception is a notion that implies success, sounds are among the causes of auditory experiences when circumstances are not in some way deviant. Happenings in alien realms do not cause normal auditory experiences, so the everyday world that our bodies, our ears, and we inhabit is well populated with sounds according to sonic realism.

Sonic realism has much to recommend it. Not only is it common sense, it respects naturalism in explaining experiences, their contents, and their causes. If perception involves responsiveness to a reality beyond the boundaries of the mind, to a world that empirical science attempts to explain, and if sounds are among the things we perceive, then sounds must have a place in the natural world. And if the natural world has no place for the sounds as we experience them to be, then our understanding of the function of audition will need revision. For it is part of our understanding of audition's function that we learn about the world we inhabit in part by perceiving the sounds it contains. Unless philosophical considerations force us ultimately to conclude that the world contains no sounds, or that sounds differ greatly from how we experience them to be, realism remains the starting point for a theory of sounds and their perception.

What follows is an articulation and defense of sonic realism. The task of realism about sounds is to provide a compelling account of the nature of sounds, and of their place in space and time. It is to convincingly answer the question, what is a sound?

Sounds, I propose, are *events*. My proposal aims to capture not only the sense in which sounds seem located in one's environment, but also the sense in which sounds are creatures of time. Much as visible colors are bound up with the extended spaces of volumes and surfaces, sounds are inextricably connected to time and have durations. A sound has a beginning, a middle, and usually an end. The sounds we hear exhibit distinctive patterns of change to their audible attributes through time. Such characteristic patterns of change determine the identities of sounds that comprise spoken words and music. Environmental sounds such as those of squirrel chatter, a falling redwood, or a loon's call all are recognizable in virtue of their unique temporal characteristics. Indeed, none could be the sound that it is without displaying such behavior through time. Since objects as we experience and conceive of them are intuitively wholly present at each moment at which they exist, and since properties or qualities do not inhabit time in the way that sounds do, I argue that events are poised as the best candidates for a theory of sounds. Sounds, according to the account I develop in this book, are happenings or occurrences that take place in an environment populated with objects and interacting bodies.

Realism about sounds, we shall see, discloses greater variety among the objects of perception than the traditional lines of thought imagine. But it leaves open a question about the richness of audition's perceptual content. Though sounds and audible qualities surely count among the objects of audition, hearing also perceptually informs us about an environment filled with ordinary objects, creatures, and happenings. How much does being so informed depend on what might properly be called hearing, and how much depends on prior knowledge or some cognitive process akin to inference? There is all the difference in the world between making an inference and enjoying perceptual awareness. Although my focus throughout most of this book is on the nature of sounds themselves—considered as immediate objects of auditory perception—the content of audition encompasses more than just those particulars that deserve the name 'sound'. The key to understanding how this might be so, I suggest in this book's final chapter, is to take seriously the import of a number of surprising *intermodal* or *cross-modal* illusions.

Cross-modal illusions, such as the ventriloquist illusion, the McGurk effect, and the recently discovered sound-induced flash illusion, all involve an illusory change to one's experience in a given modality produced by what one perceives through another modality. Such illusions, I argue, present a barrier to theorizing about any perceptual modality entirely in isolation or to treating any modality as an autonomous domain of philosophical inquiry. Explaining the cross-modal illusions recommends a dimension of perceptual content that cuts across the modalities. Most importantly, confronting the interactions among sense modalities challenges a naïvely visuocentric approach to theorizing about perception and its objects. I shall argue, at the conclusion of this book, that perhaps the greatest lesson of taking sounds seriously is one we might not have expected at the outset. The lesson for theorizing about perception is that what is most distinctive and noteworthy about perceiving—our capacity for a kind of access to the world of things and events—emerges most clearly when we consider the relationships and interactions among perceptual modalities. Thinking of each modality in isolation is valuable insofar as it informs an integrative approach to theorizing about perceptual experience. Sounds and hearing form one critical component to such an approach.

I begin by asking, simply, what kind of thing is a sound?

2

What Is a Sound?

2.1 What Kind of Thing Is a Sound?

Sounds are public objects of auditory perception. By 'object' I mean only that which is perceived—that which is available for attention, thought, and demonstrative reference. Two listeners in a room may hear and talk about the same sound, and all in attendance may hear the sound of the same speech. Though you might hallucinate a sound, in that case you fail to hear a sound. It just seems to you that you do. There is thus a distinction between genuinely hearing or perceiving a sound and enjoying an auditory experience, since it is possible to have an auditory experience without perceiving anything at all. Tinnitus sufferers suffer from auditory hallucinations, as do some who seem to hear voices.

Furthermore, if we successfully hear anything at all, we hear a sound. Whatever else we hear, such as ordinary objects or happenings in the environment, we hear by way of or in virtue of hearing the sounds it makes. So sounds are in this relatively innocuous sense the immediate objects of auditory perception. The sense of 'immediate' is innocuous because it does not imply that nothing causally intervenes or mediates our perception of sounds, and it is neutral on the question whether we are, in another sense, immediately aware of auditory sense data or of representations. I mean that when we reflect upon our auditory experience we find that being auditorily aware of anything requires being aware of a sound. But being aware of a sound requires awareness of nothing more than that sound and its attributes.[1]

Sounds frequently are characterized by the pitch, timbre, and loudness they seem to have. But this still tells us very little about what kind of thing a sound is—what ontological category it belongs to. That is a question any theory of sounds must answer.

From the outset, two initial kinds of constraint bear on the theory of sounds. The first is phenomenological. Given that sounds are among the things we hear, how we hear them to be is relevant, prima facie, to theorizing about what sounds are. All else equal, an account that captures the phenomenology of auditory perception is preferable to one that does not. The reason is straightforward. Sounds, as we hear them, seem to have certain features. All else equal, the best explanation for why audition presents sounds to have a given feature is that sounds have that feature. Presumably no such pressing phenomenological constraint pertains to theorizing about electrons, for example.

The second set of constraints is epistemological. Given that hearing grounds beliefs about sounds, ordinary objects, and happenings, an account of sounds and their perception should make possible such beliefs. The account therefore should be compatible with our having some form of perceptual access to the sounds. But, since audition informs us about the world beyond sounds, sounds also must bear to ordinary things and happenings relationships appropriate to furnishing such information. It is natural to hope that the phenomenology of auditory perception will not be at crossed purposes to an adequate epistemology of auditory belief. We might hope, therefore, that according to our account sounds audibly bear information consonant with auditory beliefs.

[1] Compare the discussion of Snowdon (1992), who claims that the immediate object of perception is just what a subject may demonstratively identify.

2.2 Sounds as Properties

The traditional philosophical outlook has, I noted earlier, grouped sounds with the colors, tastes, smells, and other secondary qualities or sensible attributes. Popular analyses of such qualities or attributes then imply that sounds are dispositions to cause auditory experiences in suitably equipped perceivers under the right sorts of circumstances, categorical bases of such dispositions, physical properties, manifest primitive or simple properties, or mere projections of qualities of experiences. The options are familiar from the literature on color.[2]

John Locke, for one, thought sounds were secondary qualities: powers grounded in the primary qualities of bodies to produce auditory experiences (Locke 1689/1975: II, viii, 10). To which *bodies* did Locke mean to attribute such powers? He may have meant to attribute them to ordinary sounding objects—bells and whistles—so that sounds, like colors, are dispositions objects have to affect perceivers' experiences (see, in particular, II, viii, 9–14).

> What I have said concerning *Colours* and *Smells*, may be understood also of *Tastes* and *Sounds, and other the like sensible Qualities*; which, whatever reality we, by mistake, attribute to them, are in truth nothing in the Objects themselves, but Powers to produce various Sensations in us, and *depend on those primary Qualities, viz.* Bulk, Figure, Texture, and Motion of parts; as I have said. (ibid. II, viii, 14)

Locke may have meant, however, to attribute sounds to the *medium* that intervenes between object and perceiver so that sounds are dispositions of the medium itself, considered as a body, to produce auditory experiences. Though this reading does not seem in the spirit of the relevant chapter of the *Essay*, in Locke's later *Elements of Natural Philosophy*, he remarks:

[2] See, for instance, the collection of papers in Byrne and Hilbert (1997*a*) for a sampling of representative views.

That which is conveyed into the brain by the ear is called sound; though, in truth, till it come to reach and affect the perceptive part, it be nothing but motion. The motion, which produces in us the perception of sound, is a vibration of the air, caused by an exceeding short, but quick, tremulous motion of the body, from which it is propagated; and therefore we consider and denominate them as bodies sounding.[3] (Locke 1823: III, 326)

Depending on whether Locke meant to attribute sounds to objects or to the medium, we get two views that differ on where sounds are located.

Robert Pasnau (1999, 2000) takes a stand on this issue concerning the locations of sounds. Pasnau introduces a view according to which sounds are properties of what we ordinarily take to be the sounding objects and not of the medium. Sources themselves *have* or *possess* sounds on this view. For Pasnau, sounds either are identical with or supervene upon the vibrations of the objects we ordinarily count as sound sources, so sounds are *properties* that depend upon the categorical bases of Lockean powers. Pasnau and Locke thus both reflect the traditional understanding of sounds as secondary qualities or sensible attributes. We can classify views developed in the spirit of the traditional model of sounds as sensible attributes according to their stance on two questions. (1) What is the correct account of the sensible qualities in general? That is, are they dispositions, physical properties, or primitive properties with which perceptual experience acquaints us? (2) Are sounds properties of the medium or of the objects? A matrix of property views of sound is the result.

However, independent of providing the details of a philosophical account of sounds as secondary qualities, we need to ask whether the sensible attribute model is the right approach to sounds in the first place. I want in what follows to suggest that it is not. Both questions that yield the matrix of property views depend upon a misguided supposition. The suggestion that sounds themselves are sensible qualities is attractive only if we are in a mood

[3] Thanks to Matthew Stuart for bringing this passage to my attention.

that overemphasizes similarities with color and entices us to provide an account that subsumes sounds with colors under a single metaphysical category.

This is because sounds themselves are not *properties* or *qualities* at all. Sounds, I claim, are *particular individuals* that possess the audible qualities of pitch, timbre, and loudness, possibly along with other inaudible properties. They enjoy lifetimes and bear similarity and difference relations to each other based on the complexes of audible qualities they instantiate. Sound sources, among which we count ordinary objects and events such as bells, whistles, and collisions, stand in causal relations of making or producing sounds, but are not at intervals simply qualified by their sounds.

This intuitive philosophical picture finds empirical support from research on audition. According to our best understanding of the central task of auditory perceiving, sounds are the particulars that ground the grouping and binding of audible qualities.

Perceiving sounds requires discerning coherent and significant streams of auditory information from an intertwined set of signals bound up with irrelevant 'noise'. Audition researcher Albert Bregman (1990: 3–6) likens this problem, which he calls *auditory scene analysis*, to determining the number, size, location, and identity of items on a lake by observing just the motions of a pair of handkerchiefs moved by the waves that travel up two narrow channels dug at the lake's edge, as in Figure 1. Right now I am able to discern the shrill sound of a small dog barking, the groan of a bus passing, the ticking of my clock, and a man's loud voice outside. The waves at my ears entwine information about these sounds into a knotty web of pressure differences.

Hearing, as we experience it, is made possible in information-rich environments by the auditory system's ability to sort through the complex information available at the ears and to extract cues about significant items the environment contains. The experienced result is a set of distinct temporally extended sounds heard as generated in the surrounding space. From pressure differences we discern the sounds of the breaking and of the sweeping. Audition accomplishes

Figure 1. Auditory scene analysis

this by grouping or bundling audible qualities into distinct auditory perceptual 'objects' or 'streams'. A set of grouping principles that involves assumptions about the objects of auditory perception enables us to associate correctly the *low pitch* with the *soft volume* and *faraway location*, and at the same time to group correctly the *high pitch* with the *loud volume* and *nearby location*, without mixing things up into a garbled 'sound soup' of *high pitch, nearness, soft volume, low pitch, loud volume,* and *distance*. Our ability to group correctly the qualities of auditory perceptual objects or streams grounds our ability to discern complex individual sounds in the environment on the basis of information arriving at the ears. Auditory scene analysis amounts to sound perception precisely because the auditory system invokes principles founded upon assumptions that capture genuine regularities in the world of sounds.

The auditory system answers the problem of auditory scene analysis by segregating the auditory scene into separate sound objects or streams characterized by complexes of pitch, timbre, loudness, and location properties. This answer, in effect, turns on the auditory system's treating the auditory objects or streams in question as particulars. First, auditory objects or streams bear complexes of pitch, timbre, and loudness, and serve as the primary locus for audible property binding. Color, by contrast, is a characteristic of particulars, not a locus for further visible attributes. Second, discrete auditory objects may be represented as distinct both at a single time and across time. That is, distinct sounds can be heard as simultaneous or successive, but qualitatively similar sounds need not be heard as identical. Third, as the term 'stream' indicates, the objects of auditory experience last through time and persist by having duration. Finally, auditory perceptual objects or streams regularly survive changes to their properties through time, as the sound of a spoken word or waning siren demonstrates. These considerations strongly indicate that auditory objects or streams are particulars that ground the grouping and binding of audible properties. Awareness of a particular auditory object or stream constitutes awareness of a sound. Sounds are audible particulars.

One might object that what researchers call auditory objects or streams are just attributions of properties to the ordinary objects we commonly think of as sound sources. To these everyday objects, we ascribe the audible property complexes that result from correct grouping. The sounds, according to this objection, are the audible property complexes. Ordinary objects serve as the particulars that ground audible property grouping.

This suggestion is prima facie plausible because we nearly always experience sounds as sounds of something: we hear the sound of the piano or the sound of the door closing or the sound of a car starting in the driveway. We might say that guesses or assumptions about auditory objects or streams are guesses or assumptions about ordinary objects like bells, whistles, and car doors. Anecdotal evidence tells both for and against the suggestion. It makes some sense to say that the bus is silver and loud. Perhaps it makes less sense to say that the very same thing has the properties of being red and being high-pitched.

Fortunately, we need not rely on anecdotal evidence. The suggestion that auditory objects or streams are just ordinary objects that possess sounds qua properties is unsatisfactory because audition is capable of representing, without error, distinct audible particulars where only a single ordinary object exists. Subjects reliably segregate distinct auditory objects or streams that come from a common object, ascribe to these auditory streams distinct audible qualities, and count the streams as distinct sounds. So, for example, you might hear the very same vehicle produce a humming sound while simultaneously generating a knocking rattle, or hear both a high-pitched whine and the sound of a voice produced by a loudspeaker. Distinct audible particulars; single ordinary object.

Bregman (1990: Chapter 3) illustrates that a single object can produce one sound, and continue to generate it while introducing another distinct sound that overlaps the first in time. Consider two contrasting cases. In the first condition, a sound gradually increases and then decreases in loudness; in the second condition, the loudness abruptly increases, remains constant for a few moments and then drops abruptly away. The abrupt change in loudness appears as the introduction of a second, simultaneous but qualitatively similar sound, while the gradual shift in loudness is taken as a change to the current sound. In the second condition, instead of a single sound that changes, a new *distinct* sound that overlaps the first in time seems to be introduced in the same place as the first, without the introduction of a distinct sounding object. Audition represents a new particular with a distinct set of audible qualities that is produced by the very same ordinary object.

In this and similar experiments, subjects do not report that the number of objects has changed and are not inclined to think that a new sounding object has been introduced at the same place where one already exists. They do, however, report the existence of a new sound. That is, subjects do not judge that audible properties already present undergo rapid change, but instead regard a distinct particular as coming into existence. More to the point, auditory experience need not incline one to think that a novel object has come into existence or started sounding. Rather, a new sound seems present, and may be generated by the same object that has been present throughout. The auditory experience may itself be neutral as to the presence of ordinary objects in such scenarios, though it clearly requires the assumption of a new auditory particular to deliver the auditory experience of a distinct sound object or stream. That is because the distinct *groupings* of audible qualities cannot be explained by reference to the very same ordinary object taken alone. An object could perhaps possess multiple sounds at a time, but how does the high pitch get bound with the loud volume, while the low pitch gets bound with the soft volume if objects are the particulars that possess audible qualities? The audible qualities must be attributed to independent particulars to explain the successful perceptual individuation of sounds.

Subjects may, nonetheless, assume that a new source corresponds to each new sound even if a new object does not. So, perhaps sounds are properties of sources, where sources are something other than ordinary objects that produce sounds. Perhaps the sources are, instead, events or happenings of some sort, and include occurrences such as collisions of toys or scatchings of fingernails on chalkboards. Similar considerations tell against this suggestion: a collision might itself have several distinct sounds, as might the strumming of a guitar. Nonetheless, source events of appropriate grain that match distinct sounds in number might yet be discerned or discovered. Evidence similar to that discussed above does not rule out such a view. But my rejection of the identification of auditory perceptual objects or streams with properties of ordinary objects or events is intended to point out that we do not regard sounds merely as repeatables that account for the dimensions of similarity among other items. Rather, sounds appear auditorily as distinct particulars that bear similarity and difference relations to each other based on their complexes of audible qualities—the properties of pitch, timbre, and loudness—to which their identities are tied. Sounds, I want to suggest, have identity, individuation, and persistence conditions that require us to distinguish them from properties of the sources that we should understand to make or produce sounds.

The identification of sounds with properties has a defect that in my view cannot satisfactorily be repaired. The defect is that it fails to account for the essential temporal characteristics of sounds. Property bearers may instantiate and persist through the loss and gain of properties and qualities, while properties, qualities, and their instances exhibit quite different temporal characteristics. This is the most convincing indication that sounds are not just properties that objects, events, or sources gain and lose.

In short, sounds have durations. A sound has a beginning, a middle, and an end. But not only do sounds continue through time, sounds also survive changes to their properties across time. A pitch shift is not the end of a sound. Unlike the way that a wall loses one color and gains another when it is painted, an object does not lose its sound and gain an entirely new one when it goes from being low-pitched to being high-pitched, as with an ambulance siren's wail. A sound can have a low-pitched part and a high-pitched part, and this is not just a matter of some source's being low-pitched at one time and high-pitched at another. Rather, a distinct particular survives the change. The orange color of the car's exterior does not survive the priming, and the dingy smell of its interior does not survive the pine scenting. The sound of the cracked muffler does, however, survive the increase in loudness and pitch upon acceleration.

It is a central fact about sounds and hearing that the identities of many recognizable sounds are tied to the pattern of audible qualities they exhibit over time. To *be* the sound of a duck's quack or the sound of a spoken syllable requires a certain complex pattern of changes in pitch, timbre, and loudness over time. The sound of the spoken word 'forest' differs from the sound of the spoken word 'foreign' precisely because the corresponding patterns of change in audible qualities differ over time. That hearing and speech researchers have spoken of sounds as *streams* reveals the significance of the role of time in understanding sounds and their perception. The auditory scene has a thoroughly temporal horizon.

This means that if sounds were properties at all, they would need to be complex structural properties characterized not just at a single time, but over many moments. And the particulars to which audition attributed sounds would have to share their existence conditions with those of the instances of sounds. Finally, the sounds would need to be capable of causing and being caused. It is, therefore, the persistence and duration of sounds that distinguishes them most sharply from the traditional secondary quality understanding implicit in much philosophical work on sensation and perception. Once appreciated, the temporal characteristics of sounds present the greatest theoretical obstacle to a phenomenologically plausible and perceptually tractable account of sounds along the contours of the property model.

All of this is not to say that no account of properties could make sense of the particularity and temporal character of sounds in a way that dealt with auditory grouping and binding through time. A trope theorist, for example, might capture the particularity of sounds not in terms of the source but by understanding sounds as particularized complexes of particularized pitch, timbre, and loudness complexes bearing particularized temporal relations to each other. The success of the theory of sounds, however, should not rest on such a controversial theory about the metaphysics of properties. My claim is that given the particularity of sounds, which is required to capture how we perceptually individuate sounds, and given the temporal characteristics of sounds, including duration and change, the property model assumed both by traditional secondary quality views of sounds and by Pasnau's more recent account of sounds as sensible physical attributes is ill-suited as a perceptually realistic candidate for an account of the metaphysics of sounds. Abandoning that model of sounds and their perception frees us from a host of cumbersome and weakly motivated metaphysical commitments. This points the way to a richer and more nuanced understanding of the objects of auditory perception.

2.3 Sounds as Waves

The standard philosophical understanding of sounds, of which I have been critical, has not gained a wide audience. Acoustic science has taught that sounds are waves. We learn early on that sounds are longitudinal pressure waves that travel from a source to our ears and that these waves are the proximal causes of auditory experiences. The sound just is the wave train leading from source to subject.[4]

Just what the customary wave view of sounds amounts to metaphysically is somewhat obscure. One way to characterize the wave is as a pattern of pressures at each point in the surrounding medium over time. This interpretation makes the wave a complex property of the medium that evolves through time. On the version of the secondary quality view that ascribes sounds to the medium, pressure patterns are candidates for the categorical bases of dispositions to produce auditory experiences. This proposal, however, is a version of the property understanding of sounds and faces just the problems that stem from treating sounds as repeatable properties instead of particular individuals. As an account of the metaphysics of sounds it makes little headway. There are, however, other promising ways to develop the view that sounds are waves. If the wave view is

[4] Sometimes the story in school is more nuanced. Consider the following description from a NASA research center educational website:

> Air is a gas, and a very important property of any gas is the *speed of sound* through the gas. Why are we interested in the speed of sound? The speed of '*sound*' is actually the speed of transmission of a small disturbance through a medium. *Sound* itself is a sensation created in the human brain in response to sensory inputs from the inner ear. (We won't comment on the old 'tree falling in a forest' discussion!) (http://www.grc.nasa.gov/WWW/K-12/airplane/sound.html)

The claim that sound is a sensation is one I reject. My interest here is realist views that do not attribute the kind of massive projective error endorsed, for example, by Boghossian and Velleman (1989, 1991) and Hardin (1993) in the case of color experience.

plausible as a view about what sounds are, then *the wave* in question is a *particular* that *persists* and *travels* through the medium.

First, waves stand in causal relations. Waves are produced or generated by their sources. Sound waves are the causal by-products of the activities of objects and interacting bodies and have among their effects the motions of resonating bodies and the auditory experiences of hearers.

Second, the wave bundle responsible for the experience also has spatial boundaries. It is characterized by a wavefront that propagates outward from the source, and its spatial extent depends on when the wave-generating activity ceases and the last pressure disturbance brings up the rear. Even when the wave train rebounds from a reflecting surface, its spatial boundaries may remain intact, even if altered or distorted. Furthermore, these spatial boundaries are perceptually significant. For example, the onset of periodic pressure differences at one ear is assumed to share a cause with their onset at the other ear, despite a delay. The shared spatial boundary responsible for differential onset is critical for auditory localization.

Third, the waves propagate or travel at a speed determined by the density and elasticity of the medium. In 20°C air at sea level, we say that the speed of sound is 344 meters per second (1497 meters per second in water; 6420 meters per second in aluminum).

Finally, waves are capable of surviving changes to their shape and to other properties and qualities. A wave's form and amplitude may change as it propagates, resulting in different heard attributes, but the wave persists throughout.

Such spatially bounded, traveling particulars are in certain respects surprisingly object-like. They can be created; they have reasonably defined spatial boundaries but persist through deformation; they survive changes to their locations and other properties; and they are publicly perceptible. To be sure, they make peculiar sorts of objects: their capacity to overlap and pass through themselves makes them stranger than most everyday objects. Though this may be a mereologically interesting problem, it seems to pose no fundamental obstacle to viewing wave bundles as in some, perhaps minimal, sense object-like.

Another important qualification to this object-like nature is that waves are dependent particulars. Sound waves depend for their existence on a medium. Their survival conditions differ from those of the medium, and they depend on different bits of the medium at different times, but without an elastic medium no sound waves exist. It is likely that lots of other things are dependent particulars, too, like tables and chairs and anything else not identical with its constituting matter. This seems to pose no obstacle to viewing the waves as object-like.

The dependence of waves on a medium is significant for a different reason. In light of the awkward fit of understanding waves as object-like particulars, the dependence points to an alternative take on the wave bundle altogether. The wave is in an important sense something that *happens to* the medium. The wave is not just a parasitic item passing through the medium; it constitutes a dynamic occurrence that takes place within the medium. The existence, propagation, and boundaries of the wave depend on processes that occur in and essentially involve a medium, so to highlight the medium dependence of the wave and its attributes is to highlight the wave's event-like characteristics. It is more plausible to think of *the waves* the wave conception of sound identifies as the particular sounds not as the object-like bundle, but instead as a variety of event that takes place and evolves in the medium through time.

Whether or not the wave view of sounds can accommodate it, the event-like construal is far more plausible as an account of *sounds* than the object-like construal. Features central to how we conceive of object-like particulars, in contrast to time-taking particulars like happenings and events, make for poor characterizations of sounds. One telling point already played a key role in rejecting the property understanding and delivers a central desideratum in theorizing about sounds. An account of sounds should capture the fact that the qualitative profile of a sound over time is crucial to its being the sound that it is, as we recognize in the difference between the sounds of 'protect' and 'protean'. But it is an intuitive feature of the way we perceive and perceptually understand objects that they persist by enduring through time, as opposed to perduring by having numerically distinct temporal parts at different times. That is, we intuitively think of objects, as opposed to time-taking particulars, as being wholly present at each time at which they exist. Intuitively, all that is required to be the desk is before me. That is what led Judith Jarvis Thomson to say of perdurantism, 'It seems to me a crazy metaphysic—obviously false' (Thomson 1983: 210). And that is why the perdurantist must motivate the view with philosophical considerations. In fact, evidence from empirical psychology may soon support an endurantist conception of the contents of object perception.[5] This fact about the way that objects appear to persist does not apply to events and other time-taking particulars, which intuitively have parts that exist and take place at different times. The soccer match, for instance, takes ninety minutes. In particular, it does not apply to sounds as we perceptually individuate them, since sounds simply are not candidates for being entirely present at a given moment. Sounds, instead, are things that occur over time.

Now, if objects do perdure, in contrast to the intuitive way we perceive and understand them, then the difference between events and time-taking particulars and objects may be just a matter of degree. In that case sounds are quite a distance from the end of the continuum occupied by tables, chairs, and even persons. In any case, I do not want my account of the metaphysics of sounds to hinge essentially on a discussion of how objects persist. What is clear is that sounds are in important respects different from ordinary objects in their ways of extending through time.

My goal in this section has been to point out that the widely accepted wave view is not completely clear either from a metaphysical standpoint or as a theory of sounds. The understanding of

[5] See Feldman and Tremoulet (2006) for a useful review and discussion of relevant results.

waves as event-like particulars is the most promising way to develop the view that sounds are longitudinal compression waves. That work seems worthwhile because the view that waves are dependent, spatially bounded, event-like particulars that persist and travel from their sources outward through the surrounding medium captures many of our commonly held beliefs about sounds. But as we will see in the next chapter, the science-inspired model of sounds as waves, like the traditional philosophical model of sounds as properties, qualities, or attributes, has important shortcomings that make it unsuitable as a philosophical account of sounds.

It is a strength of the wave view that it counts sounds as event-like particulars that persist through time. But any realist account of sounds should say not only what ontological kind sounds belong to, but also just where in time and space sounds exist. The wave account's problems stem primarily from its implication that such particulars exist or occur in different parts of the medium as time passes. The wave theorist cannot avoid this consequence. Nonetheless, the claim that sounds travel turns out to be an unnecessary and, indeed, an undesirable commitment for a theory of sounds.

3

The Locations of Sounds

3.1 Where Are Sounds?

When you hear the sound of a car driving by on the street outside your window, you learn not only whether the car has a hole in its muffler or has squealing brakes. You also learn something about the location of the car because hearing furnishes information about the locations of its objects. By listening, you learn not only about the character of the things and happenings around you, but also about where they are in the surrounding environment.

Hearing frequently prompts a grasp on the spatial arrangement of sound sources in one's environment. This, however, leaves open how audition grounds locational beliefs. Beliefs about the locations of sound sources might require 'working out'. They might stem from inferences based on auditory data, they might involve detecting changes to audible qualities, or they might demand correlation or appeal to spatial features discerned with other sense modalities. This, at least, is the dominant philosophical view spurred by Strawson's influential claim that auditory experience, in contrast to visual and tactile-kinaesthetic experience, is not 'intrinsically spatial' (Strawson 1959: 65–6). Nonetheless, according to an attractive alternative, audition itself furnishes awareness of space and locations. If audition has spatial content of its own, then audition-based beliefs about the locations of sound sources are perceptual beliefs based upon the spatial aspects of auditory experience.

Hearing, I want to argue, is spatial. Audition, that is, affords awareness of places and spatial features of one's environment. Furthermore, hearing furnishes information about the locations of objects and events in the surrounding environment by presenting sounds as located. Frequently we do hear the locations of sounds, and in hearing the locations of sounds, we hear and learn about the locations of sound-producing sources. My claim is that sounds themselves, as we experience them in audition, seem to be located not only in a particular direction, but also at some distance. Sounds, however, do not seem to travel. Sounds ordinarily seem to have stable distal locations relative to their sources. This feature of auditory experience impacts the metaphysical theory of sounds.

I argued in the previous chapter that sounds are particulars and not properties, as the traditional sensible quality view of sounds understands them. I also claimed that sounds are event-like particulars and not object-like particulars. My goal in this chapter is to establish that if auditory experience is not systematically illusory along two critical dimensions, then sounds are located at or near their sources. The widely accepted wave understanding of sounds implies, however, that sounds exist in the medium that intervenes between sources and perceivers, and it implies that sounds travel through or take place in different parts of the medium over time. The wave account of sounds, I shall argue, therefore implies that auditory experience involves two radical illusions: an illusion of location, and, perhaps surprisingly, an illusion of duration. If audition is not systematically illusory with respect to the experienced spatial locations of sounds, and if sounds have roughly the durations and temporal characteristics they seem in hearing to have, then sounds are not events of waves passing through the medium. The wave view of sounds therefore lacks the resources to satisfy the constraints on a phenomenologically adequate account of the objects of auditory perception. Unless auditory experience involves pervasive illusion with respect to both the locations and durations of sounds, our theory should locate the sounds where they perceptually seem to us to be—at or near their sources.[1]

3.2 Locational Hearing

The central fact in this discussion is that hearing, like vision and probably unlike olfaction, is a locational mode of perceiving. We learn on the basis of audition about the locations of things and happenings in the environment thanks to the spatial aspects of auditory experience. When you hear the sound of a car passing or the sound of a plate breaking on the kitchen floor behind you, you hear not just something about what is going on, but also something about where it occurs. Though hearing lacks vision's rich spatial detail, audition consciously presents information about the relative locations of audible events and objects. In *Spatial Hearing*, Jens Blauert says:

> Research has shown that the region of most precise spatial hearing lies in, or close to, the forward direction and that, within this region, a lateral displacement of the sound source most easily leads to a change in the position of the auditory event... The spatial resolution limit of the auditory system [about 1° of arc] is, then, about two orders of magnitude less than that of the visual system, which is capable of distinguishing changes of angle of less than one minute of arc. (Blauert 1997: 38–9)

The spatial information conveyed in audition, however, does not consist solely in directional information. Concerning what he calls *distance hearing*, Blauert reports:

> For familiar signals such as human speech at its normal loudness, the distance of the auditory event corresponds quite well to that of the sound source. (ibid. 45)

Blauert notes that even though distance localization is much less accurate for unfamiliar sounds, including 'unusual types of speech',

[1] Robert Pasnau (1999) first recognized that the phenomenology of locatedness, if veridical, is incompatible with the view that sounds are waves that travel through the medium.

even in such cases 'The auditory event is, to be sure, precisely spatially located' (ibid. 45–6). This is representative of the intuitive and widely accepted view among auditory researchers that hearing perceptually informs subjects about the locations of things and events in egocentric space.[2]

3.3 Located Sounds

Hearing, I shall argue, consciously furnishes information about the locations of objects and events in the environment *by* or *in* consciously presenting sounds themselves as located in the surrounding environment. You hear the car passing or hear the plate breaking behind you by hearing its sound to have some location in the surrounding space. The experience presents *the sound* as occurring at some distance and in a particular direction. You turn to look at the passing jalopy or to see how much of a mess there is because, in a wide range of cases, sounds seem to have more or less determinate locations outside the heads of their perceivers. Stanley Gelfand refers to this phenomenon as *extracranial localization*: 'Sounds heard in a sound field seem to be localized in the environment' (Gelfand 1998: 374). Hartmann and Wittenberg put it like this: 'Listeners perceive the sounds of the real world to be externalized. The sound images are compact and correctly located in space' (Hartmann and Wittenberg 1996: 3678). A contrast with the case of listening with headphones illustrates that ordinary hearing presents sounds not just as in some direction but also as at a distance. Headphone listening differs from ordinary hearing in that sounds seem to come from somewhere between the subject's ears, as in Figure 2, and not from the environment. Gelfand describes headphone listening as involving *intracranial lateralization*: 'Sounds presented through a pair of earphones are perceived to come from within the head, and their source appears to be lateralized along a

[2] See, for example, the work of Barbara Shinn-Cunningham, including (2001*a*, *b*; 2004; 2005) and Colburn et al. (2006); Bregman (1990); Hartmann and Wittenberg (1996); and the classic discussion in Mills (1972).

plane between the two ears' (Gelfand 1998: 374). Sound engineers make use of facts about localization and externalization to shape the experience of sounds in concert halls, movie theaters, and living rooms. Stereo speakers have good sound 'imaging' when they create a discernible soundstage that accurately conveys the spatial arrangement of instruments and vocalists. Audible objects and events have audible locations because their sounds have audible locations.

That sounds are heard to be roughly where the events that cause them take place is empirically supported, introspectively discernible, and sometimes revealed in ordinary language. A recent police tip sheet entitled, 'How to Be a Good Witness' instructs individuals to 'Look in the direction of the sound—make a mental note of persons or vehicles in that area' (Kershaw 2002). My phenomenological claim is that we experience sounds, in a wide range of central cases, to be located in the neighborhood of their sources. When we do not, as when a sound seems to 'fill a room' or to 'engulf' us, this is not a matter of the sound seeming to lack location. Rather, the sound auditorily appears to occupy some larger portion of the surrounding space or to be 'all around'. Seeming to hear a sound located 'in the head' when listening to

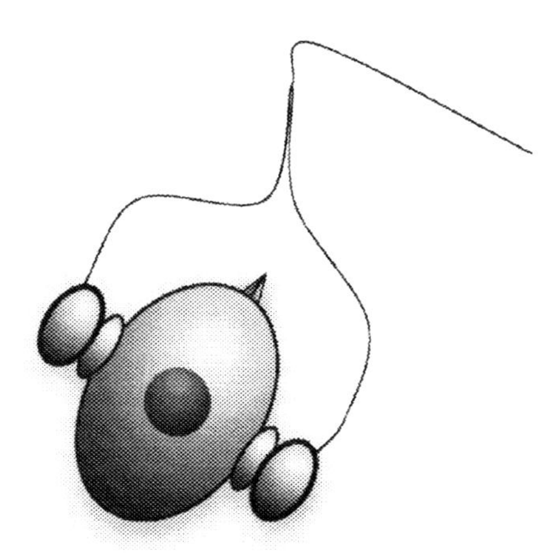

Figure 2. Headphone listening

earphones is another sort of sound location experience, though it is a bit odd.

Audition, I claim, consciously furnishes information about the locations of ordinary objects and events by presenting sounds as located. If location information is available in audition, it must be available in virtue of qualities of which one is auditorily aware. Since sounds are the immediate objects of auditory experience, however, locational information must be available, in the first instance, in virtue of awareness of the audible qualities of sounds. I shall argue in what follows that on any phenomenologically plausible account of locational hearing the audible qualities of sounds are heard to be located in the neighborhood of their apparent sources. It follows that on any such account, sounds are heard as located.

It is notoriously difficult to argue for a phenomenological claim.[3] That is what I hope to do because our account of audition's perceptual content, our understanding of audition's epistemic role, and, as I have indicated, our ontology of sounds all depend upon this claim. One way to argue for the phenomenological claim is to show that it emerges as the best among alternative descriptions of the phenomenology of locational hearing.[4]

3.4 'Coming from'

Despite the phenomenological claim that sounds seem to perceivers to be located at some distance in a particular direction, it is natural to describe sounds as 'coming from' their sources. We ask where the beeping sound is coming from and whether the hum is coming from above or below. If sounds seem to *come from* particular places, in a spatial sense of 'coming from', then locatedness as I have characterized it does not accurately capture the phenomenology of

[3] Or so an anonymous referee once told me. A more recent referee disagrees.

[4] Susanna Siegel deploys a similar method for arguing that a proposal about the content of experience is phenomenologically apt. See Siegel (2005; 2006*a*, *b*; and 2008).

spatial auditory perception. Sounds, in that case, do not seem to be located; they seem to come from particular locations.

The virtue of this way of describing the phenomenology is that it is consistent with facts about our ability to acquire information about the locations of objects and events in the environment through auditory perception. If sounds seem to come from their sources, then we can learn about where those sources are through audition without hearing sounds themselves as located just where their sources are.

How are we to take talk of sounds being heard to 'come from' a location? It might be that sounds are heard to come from a particular place by being heard first to be at that place, and then to be at successively closer intermediate locations. But this is not the case with ordinary hearing. Sounds are not heard to travel through the air as scientists have taught us that waves do. Imagine a scenario in which engineers have rigged a surround-sound speaker system to produce a sound that seems to be generated by a bell across the room. This sound subsequently seems to speed through the air toward you and to enter your head like an auditory missile, as in Figure 3. This would indeed be a strange experience unlike our ordinary experiences of sounds, which present them as stationary relative to the objects and events that are their sources.

Figure 3. Auditory missile

Perhaps, instead, sounds are heard to *be* nearby, but to have *come from* a particular place, much as a breeze seems to have come from a certain direction, as in Figure 4. But feeling a breeze is more like listening with an earphone: the tactile experience of feeling a breeze, like the auditory experience of hearing a sound played through one earphone, includes direction but no distance. Earphone listening differs from ordinary hearing not just in where sounds seem to come from, but also in where sounds are heard to be. Ordinary spatial hearing is not in this respect like tactile spatial perception. Imagine feeling *where the fan is* by feeling its breeze.

Since sounds seem to come from sources in a sense that includes distance as well as direction, and not in a sense that includes travel, neither spatial understanding of sounds' seeming to come from their sources does justice to the phenomenology of ordinary auditory experience. Thus neither spatial understanding of 'coming from' explains how auditory perception furnishes information about the locations of sound sources. The best sense to make of sounds' seeming to come from particular locations is that they have *causal sources* in those locations—that they are *produced* or *generated* at those locations.[5]

Figure 4. Auditory breeze

[5] See Nudds (2001) for more on the perceptual experience of the generation of sounds by sources.

3.5 Sounds without Locations?

I have argued that sounds seem to have stable, distal locations and that they do not seem to travel or come from their sources in any spatial sense. Part of the motivation for this line of argument is that our awareness in audition of the locations of ordinary things and happenings in the environment requires explanation. If we are aware of locations of sounds, and if sounds seem located near their sources, then the beginning of an explanation is in hand. All that remains is to explain the mechanisms concerning how sounds are localized. Empirical research on spatial hearing has provided answers based on the physics of sound waves and on our ability to determine direction and distance information primarily in virtue of differences in wave characteristics at the two ears. For instance, interaural time and level differences, head-related transfer functions, and secondary reflections all play a role in determining the spatial characteristics of one's auditory experience (see Blauert 1997).

However, might this process, which I have invoked to explain how sounds are perceived to be located, serve instead to explain why objects and events seem in audition to be located without adverting to located sounds?

A promising approach based on this idea again rejects the phenomenological claim as I have characterized it. This approach says that we hear sounds to have pitch, timbre, and loudness, though not as having location. Rather, we hear ordinary events and objects as located and as the generators or sources of audible qualities that lack spatial characteristics entirely. We simply fail to perceive the locations of sounds. Sounds on this view are not heard to have locations, they are heard to have located sources. This claim is supplemented by an explanation in terms of sound wave transmission for how we localize sources in the environment. A location is thereby ascribed to a source, though not to a sound.

This description provides an account of the phenomenology of auditory experience without claiming that sounds are heard to have locations. To see why it fails we need to consider the way in which audition furnishes perceptual information about the locations where sounds are generated.

I have claimed that hearing provides information about the ordinary objects and events that surround us—notably, information about where those things are and occur. The account we are considering is that we hear objects and events as located by means of the sounds they generate, though we do not hear those sounds as located. One's experience of non-located pitch, timbre, and loudness must therefore ground one's auditory experience of the locations of ordinary objects and events.

I claim, however, that we cannot *hear* just non-located audible qualities and located objects, *full stop*. This would amount to a precarious perceptual situation. How could hearing non-located qualities furnish perceptually manifested locational information about sound sources?

One way has already been mentioned: locational information might be encoded temporally, for example, by time delays between waves reaching the ears. The suggestion, however, is that we are auditorily *aware* of the locations of things and happenings, since hearing is spatial. This information therefore must be conveyed somehow in conscious perceptual experience. That is, the objects and events must be presented in auditory experience as located, and this must be accomplished by means of awareness of audible qualities. At a basic level of awareness, audition presents just complexes of pitch and timbre with loudness and duration, so an auditory experience that conveys information about the locations of material objects and events must do so by means of one's awareness of these basic attributes. For an experience grounded in the audible qualities to be an auditory experience of location, the audible qualities must themselves bear spatial information. Temporally encoded location information therefore must be manifested through one's experience of pitch, timbre, and loudness. Since, as I have argued, sounds and their audible qualities do not auditorily seem to come from particular locations in a sense that involves travel or arrival, auditory awareness of location must occur thanks to an awareness of located audible qualities. Sounds, the bearers of audible qualities, must appear to occupy stable distal locations if we are to learn of those locations through spatial auditory experience.

Another way to put this point is as follows. Some feature of the sound in virtue of which one hears a thing or event must ground one's experience of its spatial location. This is because, in general, one cannot experience some particular as being at a place unless one experiences some of its (audible, visible, perceptible) qualities as being located at that place. The audible qualities in virtue of which one hears things and events are, at bottom, temporally extended combinations of pitch, timbre, and loudness. Since the experience of non-located audible qualities such as pitch, timbre, and loudness cannot ground an experience of located things and events, the experience of located audible qualities must do so. But the experience of located pitch–timbre–loudness complexes just is the experience of a located sound, since sounds are the primary bearers of pitch, timbre, and loudness. The problem with the proposal being considered is that it fails to provide a coherent account of spatial hearing consistent with the mediatedness of object and event awareness in audition.

A distinction thus can be drawn between hearing sounds themselves as located and perceiving information about the locations of material objects, stuffs, and events in the environment by means of audition. Given that we learn the locations of ordinary objects and events in audition, the question is whether the latter would be possible without the former. Since sounds are the immediate objects of auditory experience, since auditory awareness of location occurs by means of awareness of audible qualities, and since sounds seem to come from their sources only in a causal sense, hearing sounds and their qualities as located is required in order to perceptually experience or form judgments about the locations of material objects and events through audition. Sounds are heard to have locations, by means of which they provide perceptual information about the locations of their sources.

Another way we might imagine the situation is as follows. We do not hear sounds as located in any way, but instead learn the locations of sound sources by an unconscious process, something like what occurs in blindsight. Location information thus is represented through a course of subconscious perception. In this case, audible qualities do not audibly bear location information, since by hypothesis location information is not experientially manifested. Subjects simply find themselves believing that sound sources have certain locations without reflective awareness of experiencing the locatedness of those sources. This story is implausible as a characterization of ordinary hearing's phenomenology. Subjects who report and act upon the audible locations of events in egocentric space neither guess nor report beliefs lacking grounds in auditory experience. It seems we hear where things are and where sounds come from. Ordinary hearing differs from the auditory analog of blindsight in that audible qualities seem to have audible locations—a fact demonstrated by behavioral and introspective tasks. If this appearance is illusory and the blindsight explanation is plausible, then we have a fresh example of reliable perceptual beliefs formed in the absence of genuine conscious awareness. This seems improbable. It is far less improbable that we are simply aware of locations in audition.

The claim that sounds phenomenologically are located in the environment at a distance in some direction grounds an important fact about locational hearing. It is clear that we gain information about the locations of items and happenings in the environment by means of audition. Furthermore, this locational information is perceptually available to us in audition—we can form defeasible beliefs about the locations of environmental constituents just on the strength of conscious auditory experience. But sounds do not seem to come from their sources in a sense that includes travel from those sources, and sounds do not seem to come from their sources in the sense that they seem only to be nearby but to have originated at their sources. (Sounds seem to come from their sources only in the sense of being produced or generated by those sources.) It therefore follows that awareness of the sounds and their audible qualities furnishes or bears locational information since they are in the first instance that of which we are auditorily aware. Hearing sounds as located makes possible our auditory awareness of the ways that everyday objects and events are arranged in the environment and grounds perceptual beliefs about such arrangements.

Strawson (1959) nonetheless claimed that because audition is not intrinsically spatial, a purely auditory experience would be entirely non-spatial. Though I am concerned primarily with ordinary spatial hearing, I believe Strawson is wrong to sharply distinguish audition from vision and tactile-kinaesthetic experience. The claim that a purely auditory experience, in contrast to a purely visual experience, would be entirely non-spatial is false if 'purely' means just 'wholly' or 'exclusively'. Even a tone presented by headphone to one ear carries some spatial information. If an exclusively visual experience could be spatial, so could an exclusively auditory experience of sufficient richness. Recent empirical research on neural representation of auditory space indicates that 'Auditory space maps can be generated without visual input, but their precision and topography depend on visual experience. So, for example, owls raised as if they were blind end up with abnormal, or even partially inverted, auditory maps' (Carr 2002: 30).

Certain impoverished experiences of just pitch and timbre, with loudness and duration, might nevertheless count as 'minimal' or 'pure' auditory experiences. And such minimal auditory experiences might be non-spatial since pitch, timbre, and loudness are not themselves intrinsically spatial attributes. However, the visual experience of a uniform gray ganzfeld, or of a bright, uniform flash of light, might count as a minimal visual experience that is non-spatial. Likewise, a uniform feeling of warmth on every surface of the body might be a minimal, though non-spatial, tactile experience. If non-spatial minimal visual and tactile-kinaesthetic experiences are possible, then the claim that the purely, in the sense of 'minimally', auditory experience differs from purely visual and tactile-kinaesthetic experiences in being non-spatial is false.

The only sense in which audition arguably is not 'intrinsically spatial' is one that involves experienced features of the primary objects of audition. Pitch, timbre, and loudness are not inherently spatial attributes, as are shape and size. In addition, sounds are not experienced as having the rich *internal* spatial structure and articulation that ordinary visually experienced objects possess. A sound does not auditorily appear to have a detailed surface with texture and sharp edges. But to say that audition fails to represent richly detailed internal spatial features of sounds does not imply that audition is non-spatial or that audition fails to represent the locations of sounds in space. Clearly, there are places in egocentric space that auditorily appear to contain sounds and there are places that do not. Audition's coarseness of spatial grain does not constitute an all-out inability to represent space or spatial characteristics of sounds, such as their locations and approximate boundaries. It also does not imply that sounds themselves lack all spatial features we are unable to discern.

3.6 Locatedness and the Metaphysics of Sounds

I have argued that sounds, as we auditorily experience them, seem located and that sounds seem to travel only if their sources do. In addition, awareness of sounds provides our conscious mode of perceptual access to the locations of things and events in the environment through audition. For these reasons, a theory of sounds should locate the immediate objects of hearing at a distance from perceivers in the neighborhood of their sources. The alternative is to ascribe systematic and pervasive illusion to auditory perception. Not only do we sometimes get the locations of sounds wrong in hearing if sounds are not distally located and relatively stationary, we never perceive a sound to occupy its

[6] See O'Callaghan (2007*b*) for a full defense against skeptics about spatial audition.

true location. If the phenomenological claim is correct, and if auditory experience is not systematically illusory with respect to the locations of its objects, then sounds do not travel through the surrounding medium. If sounds do not propagate or travel through the medium, the understanding of sounds as waves fails.

3.7 The Durations of Sounds

The illusions multiply. If sounds indeed propagate through the medium, then the illusion is not isolated to locational hearing. The wave understanding of sounds is unable to account satisfactorily for a critical dimension of both sounds and auditory experiences. It, like the property understanding, fails to capture the striking and essential temporal features of sounds.

Clearly, perceiving the durations of sounds is an important part of auditory perception. Sounds inform us about happenings in and states of our environment, and part of what they inform us about is how long those happenings and states last. I learn through hearing when the coin stops spinning, when the fridge starts up and shuts down, and how long the car idles in the driveway. I experience how long the nine-year-old who lives next door practices the violin each afternoon—I sometimes wish the sessions had shorter durations. Now, if sounds are spatially bounded particulars whose locations in the medium change from moment to moment as do those of waves, what in fact I experience when I take the sound to have duration is not the duration of a sound at all. Rather, my encounter with a *spatial* boundary of a sound leads to my enjoying an auditory experience while the sound passes. On later encountering the far boundary of the sound, I experience the sound to end. Whether the wave is an object-like particular that passes by or an event-like particular that unfolds at different places in the medium over time, domino-wise, my experience of the sound is caused by the spatial parts of the sound wave bundle as it passes. In experiencing the sound, I do not experience the lifetime of an object-like entity or the duration of an event other than my own sensing. Apparent duration perception results from encounters with the spatial boundaries of sounds, according to the wave view. This means that each time I hear a sound, I mistake an experience of the spatial boundaries of a sound for an experience of the duration and temporal boundaries of that sound. The experienced duration of a sound, therefore, is nothing more than a form of crude projective error: I mistake the duration of an experience alone for the duration of a thing I am experiencing. Duration perception is a wholesale illusion if sounds are waves.

Perhaps you are willing to live with the illusion to preserve the commonly accepted scientific view of sounds. So suppose the experience of a sound's duration is an illusion. Since experiencing a sound mediates awareness of sound-producing events, and, in particular, since experiencing the duration of a sound mediates awareness of the duration of a sound-producing event, awareness of the durations of sound-producing events is mediated by illusory awareness of the durations of sounds. We have no reason, however, to doubt that awareness of the durations of sound-producing events is veridical. Such awareness regularly grounds true perceptually based beliefs. It follows that this case constitutes an instance of veridical mediated awareness that is mediated by an illusion. It is important here to keep in mind that the mediation in question is of a sort to which the subject has access. It is not, for example, the kind of mediatedness in question when we say that hearing is mediated by activity in the cochlea or auditory nerve.

Perhaps veridical awareness mediated by an illusion is intelligible. Things are not so bad, you might think, since the illusion informs you about something you really care about: the duration of the event that produced the sound (though it should be clear by now that I think you have good reason to care about the duration of the sound itself). But consider the source of the justification for your perceptually based beliefs about the durations of sound-producing events. That justification is based on your experience of the durations of sounds, and requires no more than enjoying such an experience. Furthermore, it is plausible to think that information available to the subject of the experience is what justifies the belief. Veridically hearing the duration of a sound *could* therefore justify your belief that a corresponding sound-producing event lasted, say, thirteen seconds. If, however, your experience of the duration of the sound is an illusion, then that experience cannot *by itself* serve as your guide to the duration of a sound-producing event. Information from the experience of duration alone cannot introspectibly justify your beliefs about the duration of sound-producing events. The experience is at best a justifier in virtue of its reliable connection to the durations of sound-producing events. If such experiences justify corresponding beliefs in virtue of information available to the subject, to the experience must be added the further belief that the durations of auditory experiences typically are tied to the durations of sound-producing events. That belief cannot be justified on the basis of auditory perceptual experience alone because it is a matter of empirical discovery; it might have been, for instance, that waves in a packet did not travel at uniform speed. The claim is not that you *could not* be justified in your mediated beliefs about the durations of sound-producing events; the claim is that such beliefs cannot be held simply on the strength of information accessible in auditory experience alone.

These complications strike me as important negative consequences of a commitment to illusory sound duration perception, since it is arguably among the primary functions of auditory perception to inform us about the temporal characteristics, including the durations, of happenings in our environment. We care about how things change and how events unfold over time, and we value audition for its superior capacity thus to inform us. The account of sounds as waves entails that we do not hear the temporal features of sounds things make as those sounds unfold over time. It entails that we do not hear the durations of sounds and that our justification for believing that the violin practice lasted forty-five minutes cannot come just from hearing because what we experience in hearing is an illusion. These consequences result from the claim that sounds construed as waves travel and pass through the medium.

The other important consequence is that our ways of perceptually individuating and tracking sounds through time are wildly misguided. If sounds persist and travel in the manner of the waves, then our perceptually based estimates of the lifetimes and survival conditions of sounds all are incorrect because the waves may continue to exist long after the sound has seemed to cease. It is simply a mistake according to the wave account to state that 'Time Is on My Side' by the Rolling Stones is three minutes and one second long if the song is the sounds. I shall have more to say about the temporal features of sounds, and about the relationship between duration and travel, when I turn later to the topic of echoes.

I have claimed that the account of sounds as waves runs into problems with duration perception. The wave account makes the perceived duration of a sound, in addition to its perceived location, a systematic illusion. The problem lies in saying that the sound—what one most immediately hears—is the passing bundle of waves. Suppose we omit the claim that sounds travel. Because it is a central fact about pressure waves that they propagate or travel through a medium, we must then abandon the suggestion that sounds are waves. I contend that the illusions of location and duration warrant doing just this.

Sounds, I claim, are located roughly where we hear them to be: at or near their sources. The sound does not travel as do the waves. The waves, however, are causally intermediate between the sounds and the auditory experiences of perceivers. The waves bear or transmit information about the sound events through the medium and furnish the materials for auditory experience. Sounds are stationary relative to their sources. If sounds are stationary events, then the auditory experience of location does not involve a systematic and pervasive illusion, and audition-based beliefs about the durations of sounds are for the most part true.

4

The Argument from Vacuums

4.1 Sounds in Vacuums?

A commonly shared assumption is that vacuums hold no sounds. If the standard science-inspired view that sounds are waves that exist in and travel through a medium such as air or water is correct, then there are no sounds in vacuums and the shared assumption is true.

But in the last chapter I presented what I take to be two of the strongest arguments on behalf of the claim that sounds are not themselves located in the surrounding medium. If sounds occur in and propagate or travel through the medium, we are subject to a systematic illusion of locatedness in auditory perception. No phenomenologically plausible account that fails to locate audible sounds near their sources explains how auditory perception conveys information about the locations of everyday events and objects. I argued, in addition, that if sounds travel through the medium, then we mistake the duration of an experience for the duration of a sound perceived. This amounts to a failure to make contact in one constitutive respect with the world of sounds.

Both problems, we should note, can be avoided by the version of the secondary or sensible property account that ascribes sounds to ordinary objects. Pasnau's (1999) account, for example, makes sounds properties objects possess just in virtue of their vibrations. This locates sounds at their sources and precludes travel through the medium. But it avoids the problems at the expense of removing the medium from the account of what sounds are. Pasnau's account makes the *medium*—the air or water in which sound waves travel—a mediator to perception, but not something required for the existence of a sound. The medium bears and transmits information about sounds, but it is neither an essential component of the sound nor a condition necessary for a sound. It follows, counter to our long-tutored intuitions, that sounds can exist in vacuums. Sounds occur even when an object vibrates alone in the absence of a surrounding medium. One need not, however, endorse a secondary or sensible property account of sounds to hold that sounds occur even in a vacuum. Casati and Dokic (1994, published in French; 2005) claim that sounds are events constituted by the vibrations of objects. Sounds, according to Casati and Dokic, therefore can exist in a vacuum.

I have claimed that sounds are not properties, but that sounds are extra-mental particulars that persist through change. I have argued that sounds are best understood as particular events, that they are located at or near their sources—where we hear them to be—and that they have roughly the durations we hear them to have. This supports a natural account of audition's capacity to reveal information about the locations and temporal characteristics of things and happenings in the environment. The places and durations of sounds are reliably connected, in a way that is open to perceptual experience, with the locations, durations, and patterns of change to objects and events around us.

Locating sounds at or near their sources appears, prima facie, in sympathy with the claim that the medium is not a necessary condition on a sound's existence. However, I have claimed that sounds are produced or generated by the activities of ordinary objects and events, and so stand in causal relations to those objects and events. This perhaps supports the requirement of a medium. Under the conditions in which we ordinarily hear them, sounds involve the disturbing or setting into wavelike motion of a surrounding medium. That sort of event has to do with the way a medium is affected by the disturbing activities of everyday objects and happenings. It requires a medium for its existence. An object's vibrating is not the disturbing of a medium if no medium is present.

Are there sounds in vacuums? The question should be decided neither by simply consulting common sense nor by reading off the consequences of one's favorite metaphysical theory of sounds. I will argue, however, that even independent from explicit theoretical commitments concerning the nature of sounds, we have reason to resist the claim that there might be sounds in vacuums. This suggests that vacuums place a correctness constraint on theories of sounds, and are not mere spoils to the victor. If there can be no sounds in vacuums, a medium is not just a necessary condition on veridical sound perception; the presence of a medium is a necessary condition for the existence of a sound.

4.2 The Argument from Vacuums

In Berkeley's first dialogue between Hylas and Philonous, Hylas deploys this *argument from vacuums* in service of the claim that sounds must be in the medium:

> *Philonous.* Then as to sounds, what must we think of them: are they accidents really inherent in external bodies, or not?
>
> *Hylas.* That they inhere not in the sonorous bodies, is plain from hence; because a bell struck in the exhausted receiver of an air-pump, sends forth no sound. The air therefore must be thought the subject of sound. [The sound which exists without us] is merely a vibrative or undulatory motion in the air. (Berkeley 1713/1975*b*: 171–2; also quoted in Pasnau 1999: 321)

Since we hear no sound when a bell is struck in a vacuum, Berkeley's first premise—that there are no sounds in vacuums—finds support from the idealist dictum that nothing exists unperceived. Barring the dictum, however, we would like to have some reason for denying or affirming that sounds exist in vacuums. The problem (though no problem for Berkeley) is that the fact that no sounds ever are heard in the absence of a surrounding medium shows only that a medium is ordinarily required for there to be veridical perception of sounds. It does not show that a medium is necessary for there to be a sound. Since we hear sounds in virtue of sensitivity to sound waves, the sound of the bell in a vacuum is inaudible. But inaudibility is compatible with there being a sound that could be heard if only there were a medium around. Consider the case of colors in the dark. Objects do not shed their colors in the absence of light. The presence of light is a condition necessary for seeing the colors of things, but not for objects to have their colors.

Intuition perhaps motivates thinking that the case of sounds in vacuums is similar. Suppose we strike a tuning fork in a chamber filled with air, then quickly remove all the air with a super vacuum pump, wait a few moments, and let the air back in. We first hear the sound of the tuning fork, then hear nothing, and finally hear it again, all while watching the tuning fork vibrate as it does when making a sound.[1] Does the sound continue through the removal and replacement of the air? Intuition may suggest that it does, and that we simply fail to hear it during the vacuum stage. Perhaps sounds in this sense are like objects in the dark or like objects that pass behind barriers. Unlike objects in the dark, however, the presence of sounds in vacuums cannot readily be confirmed perceptually by extra-auditory means. And sounds, unlike colors, do not appear as properties or features that qualify the things, such as bells and whistles, that we do believe to persist in the dark or in a vacuum. An intuition that sounds persist in vacuums therefore cannot be grounded in the persistence of an unchanging object, as can belief in the persistence of colors in darkness. The case of vibrations in a vacuum also differs from that of objects that perceptually seem to persist as they pass behind barriers. A sound may be perceptually understood to persist through masking by another sound, but ordinarily a period of silence perceptually signals the conclusion of one sound and the chance for the start of another. A better analogy is the complete disappearance of

[1] Thanks to Roberto Casati for this example in his comments on my paper, 'Sounds and Events', during the University of London conference on Sounds in January 2004.

an object before one's eyes, followed by a reappearance of one that is indistinguishable. Perceptually based intuition delivers no univocal verdict about the object's continued existence during disappearance. In deciding whether sounds exist in vacuums, given the absence of independent confirmation, intuitions about whether the sound persists through the removal of the medium, especially those that conflict with how perception customarily parses sounds, seem beside the point. We want to know whether the sound is there, and not whether untutored intuitions coupled with visual evidence of vibration lead us to suspect that it is there. The interesting question is precisely the one that cannot be answered on those grounds alone. Short of Berkeleyan assumptions, the epistemic resources of perception alone are not up to the question. Adding assumptions that are not theoretically innocent fails to resolve the matter satisfactorily.

Why not think that the sound is present, though inaudible, in the vacuum? Talk of sounds and vacuums might end here, until we have chosen among competing accounts of the metaphysics of sounds. Part of the problem is that without making such an explicit theoretical commitment, it is difficult to discover decisive reasons for thinking sounds do or do not exist in vacuums. Since the correct theory of sounds is precisely what is at issue, we cannot assume a favorite and read off its verdict concerning vacuums. But since an hypothesis affirming the existence of sounds in vacuums could never auditorily be confirmed or disproved, how could vacuums hold lessons for theorizing about sounds?

Despite the experiential dilemma, we might discover that good philosophical reasons suggest that there could be no sounds in a vacuum. I believe such reasons exist. My contention is that since no suitable ascription of audible qualities to sounds is available in the case of vacuums, we should conclude that no sounds occur in vacuums. If that is right, then although the medium is not the sole subject of sound and the sound does not inhabit the medium, a sound's existence depends on the presence of a medium.

4.3 The Medium as a Necessary Condition

Pitch, timbre, and loudness are qualities first and foremost of sounds. Ascription of pitch et al. to objects or to sound sources depends upon their producing or tending to produce sounds with these qualities. Flutes are high-pitched because they are disposed or tend to produce sounds high in pitch. A flute now has a high pitch because it currently makes a high-pitched sound. Sounds produced by an object vibrating in a given way, however, are perceived to differ in audible qualities when the vibration takes place in different surrounding mediums. The sound of a tuning fork appears to have different audible qualities in water, air, and helium. The perceived audible qualities of sounds vary across different surrounding mediums, and vibrations that take place with no surrounding medium appear to have no audible qualities. What audible qualities, then, does the tuning fork vibrating in a vacuum possess?

A straightforward answer is that because the audible qualities of a sound vary relative to the medium, the tuning fork in a vacuum possesses no audible qualities. Unless we grant that there are some special sounds without pitch, timbre, or loudness, the tuning fork in a vacuum thus has no sound. The issue is not simply that we are unable to hear the sound, but that there are no audible qualities present to be heard.

Of course, the color appearances of things notoriously vary across changes to lighting. A tie might match a sock in fluorescent light but mismatch it in daylight. Are colors therefore relative to lighting, and do things thus lack colors in darkness? One reason to resist this already has been mentioned. Colors qualify their bearers, but lighting alone seems not to change the objects that possess colors. A response familiar from the case of colors avoids relativizing to lighting conditions by appealing to standard or ideal conditions which are revelatory of the true or genuine colors of things: colors that do not vary with changes to lighting; colors that remain in the dark. Applied to the case of sounds, one could say that the sound of the tuning fork in a vacuum has the loudness it has or would appear to have in air or some other privileged medium. Though air at sea level seems a natural choice for a standard condition, why not say that the tuning fork's true audible qualities are those it would appear to have in water or helium or liquid mercury?

Deploying this strategy requires ignoring that sounds do not seem to qualify their bearers, and so might change in a way that depends on circumstances beyond changes to their sources, such as changes to the medium. Nonetheless, even assuming this, if audition ever reveals to us the audible qualities of sounds, any decision on the question concerning a sound's true audible qualities across all mediums appears highly arbitrary. How could the true audible qualities of a sound be so arbitrary? It might be noted, however, that the analogous decision in the case of vision and colors also seems arbitrary. Why couldn't the true color of an apple be the color it appears to have in fluorescent illumination or the light of some very different sun? If color appearances vary with lighting conditions, if arbitrariness tells against choosing a normative standard, and, so, if colors are relative to lighting conditions, then do colors fail to exist in the dark? Apart from appeal to the fact that colors seem 'stuck on' or to qualify their objects and sounds do not, what differentiates darkness from a vacuum?

The choice of a standard or ideal condition for seeing a thing's color is not as arbitrary as it might first appear and as some have argued. Hardin (1993), for example, has argued that the notion of standard conditions is highly arbitrary. I mean here to resist Hardin's claim by showing that our choice of standard lighting conditions is not nearly so arbitrary as Hardin thinks. Color theorists therefore have a firm foothold in responding to concerns about variability in color appearances. Daylight, for instance, delivers circumstances highly advantageous for viewing the colors of things. That is because the colors of objects are intimately tied to the reflectance properties of their surfaces. The reflectance properties of interest are those in virtue of which a surface tends to absorb and reemit light energy of different wavelengths across the visible spectrum. For instance, the *surface spectral reflectance* of a surface is the percentage of incident light that surface reflects at each wavelength across the spectrum. Though perhaps not identical with colors, surface spectral reflectances are excellent physical candidates for the causal, dispositional, or constituting basis of color.[2] Reflectance properties of surfaces explain color experiences. Full spectrum illumination (which daylight approximates) therefore has *normative significance* in revealing the colors of things because when reflected it carries information about how objects tend to interact with light across the entire spectrum. Full spectrum light reveals with fidelity the color appearances of surfaces because, when reflected, it incorporates complete detail about the range of reflectance characteristics a surface exhibits. Light missing spectral constituents lacks full information about a surface's reflectance. The apple's color is the one it appears to have in full spectrum light because this variety of illumination provides maximal information about the way the apple's surface interacts with light.

No such normatively significant medium exists in the case of sounds. Neither air nor water nor helium optimally reveals the subtle vibrations of an object in the way that full spectrum light does for the reflectance characteristics of a surface. Since the apparent audible qualities of sounds vary across different mediums, since the choice of air at sea level (or at 2000 meters, or at 20°C, ...) is arbitrary, and since any of a range of different mediums is equally good for hearing sounds, we should conclude that the audible qualities of a sound depend upon the characteristics of the specific medium in which the sound occurs. The audible qualities are medium relative.

Since objects that vibrate in a vacuum are inaudible, since there is no standard that determines the medium-independent audible qualities of a sound, and since a vacuum contains no medium,

[2] See Hilbert (1987), Byrne and Hilbert (1997*b*, 2003), Tye (2000), and Bradley and Tye (2001) for physicalist accounts of colors as reflectances. See also Johnston (1992), McLaughlin (2003), and Chalmers (2006) for accounts on which colors are tied to reflectance properties.

the best thing to say about a tuning fork in a vacuum is that it simply has no audible qualities. We are justified in concluding, therefore, not just that a necessary condition on sound perception is missing, but that a condition necessary for the existence of a sound is missing in a vacuum.

4.4 Involving the Medium

If Berkeley's first premise is indeed established by considering the medium relativity of audible qualities, what are the consequences for theorizing about sounds? Unfortunately for the wave theorist of sounds, the necessity of a medium does not entail that the medium itself is the subject of the sound or that the sound exists entirely in the medium. So, even though the unmodified property theories of Locke's *Essay* and of Pasnau (1999) are inconsistent with the vacuum arguments above, they might be modified to make the presence of a medium a necessary condition. For example, the property theorist might claim that a sound is the property an object has just when it vibrates in the right way in the presence of a surrounding medium. Or it might be the disposition to produce an auditory experience when in the presence of a surrounding medium.

But the modification must in some way make the audible qualities of sounds depend upon the medium in which the sounding occurs. This diminishes the intuitive appeal of those formerly simple theories, since they no longer conceive of sounds entirely in terms of what goes on with an object. Since the properties of the medium in part determine the qualities of the sound, considering the pattern of vibrations of an object no longer provides the full story about its sound. The modification nonetheless is required. For this reason, a hybrid account is best placed to capture the truth both in the wave theories that require a medium and in the object-based theories that locate sounds near their sources. The hybrid I propose should make the sound medium involving, but not entirely in the medium, and object involving without being determined entirely by the activities of the object. Sounds should occur when vibrations or collisions take place in the presence of a medium. Sounds, according to the account I shall propose in the next chapter, are events constituted by the *interactions* of objects and bodies with the surrounding medium. Sounds are essentially a matter, in part, of the behavior of an object or objects and, in part, of the surrounding medium and its specific characteristics. Audible qualities of sounds thus are medium dependent. The medium is not the sole subject of sounds; neither is the object. To locate the sounds we must locate the place where the objects interact with the medium. There are no sounds in vacuums.

5

Sounds as Events

5.1 Sounds Are Events

Understanding sounds as waves gets several things right. According to the best version of the account, individual sounds are particulars that one can count and quantify over, and which exemplify a range of attributes, including plausible candidates for the audible qualities. For instance, since pitch is tied to frequency, and since waves possess frequency, waves arguably bear or ground pitch. Sounds, however, need not be repeatables or properties ascribed either to ordinary objects or to the medium. A wave-based account recognizes that sounds are temporally extended occurrents with temporal parts and durations, and it counts sounds as persisting particulars capable of surviving change. Under its best interpretation, sounds are event-like particulars. But the wave account is unable correctly to capture the temporal characteristics of sounds and the nature of our perceptual acquaintance with sounds that extends through time. In short, it mistakes the lifetime of a train or bundle of sound waves in an environment for the duration of a sound.

The claim that sounds are particular events nevertheless captures important truths about sounds and meets defining desiderata for a theory of sounds and the objects of auditory perception. Sounds, intuitively, are happenings that take place in one's environment. This is evident in the language we use to speak of sounds. Sounds, like explosions and concerts, *occur*, *take place*, and *last*. Colors, shapes, and fiddles do not.

Sounds also stand in causal relationships. They are caused by commonplace events such as collisions and vibrations, and they give rise, for example, to reverberant vibration, auditory experiences, and recordings. According to the standard account of causation, causal relata are events.

Though sounds appear to have clear temporal boundaries that circumscribe durations, the spatial boundaries of a sound are less obvious. Unlike objects, whose spatial boundaries are relatively clear and whose temporal boundaries are more difficult to discern, events are characterized by clear temporal boundaries but by spatial boundaries that may be unclear (Casati and Varzi 2006). Sounds, in addition, appear to tolerate colocation or overlap with other sorts of things and events. A sound might occupy part of the same region as a fiddle or a bowing. Sounds, that is, appear to relate to space and time in ways characteristic to events. Understanding sounds as events of some sort amounts to a powerful framework for a satisfactory account of both the metaphysics of sound and the contents of auditory experience.

There is one caveat. The critical features of the theory of sounds should not turn on some one account of the metaphysics of events. I would like the theory of sounds to be reasonably neutral on the nature of events and viable no matter what events turn out to be (candidates include, for instance, theories stemming from Davidson 1970, Kim 1973, Galton 1984, Lewis 1986, and Bennett 1988). One might even hold it against a theory of events if it fails to capture facts about sounds. So, within reason, whatever events turn out to be, sounds should count as events. I think there is good chance for this, though once we get down to the detailed theory of sounds, some decisions will turn on just what is the right account of events. I want for now to operate with an intuitive conception of events as potentially time-taking particulars—as happenings that may or may not essentially involve change. Events as I wish to understand them are immanent or concrete individuals located in space and time.

Sounds, among the events, are akin to processes or activities, though some sounds—such as spoken words, birds' calls, or an eighth-note at C sharp—may lend themselves to treatment as performances or accomplishments with a certain natural trajectory toward completion. Sounds are not instantaneous events, but require time to unfold. They may tolerate shared location with their sources and with other events that occur in the medium. Sounds construed as audible occurrences are poised as causal relata. They stand in causal relations to the activities of the objects and events that we call the producers or sources of sounds, and they fulfill the causal requirement on any account of their veridical perception.

I have argued up to now that sounds are non-mental, that sounds are particulars and not repeatable properties or qualities, that sounds are located near their sources but do not travel through the medium relative to those sources, that sounds have roughly the durations we hear them to have, that the medium is a necessary condition on the existence of a sound, and that sounds are events. A theory of sounds must, in addition, identify the sounds among the events.

5.2 Disturbings

Which events are the sounds? Consider the case of a tuning fork struck in air. The striking of the fork makes or causes a sound in virtue of the oscillating arms of the fork disturbing the surrounding air and creating regular compressions and rarefactions. However, since sounds do not travel through the medium but remain stationary relative to their sources, the sound does not travel as do the waves. Since sound waves that reach the eardrums cause auditory experiences, sounds must be causally intermediate between ordinary, everyday events and traveling sound waves. Since waves bear and transmit information about sounds, sounds cause waves. And since sounds indicate something about the events and happenings that occur in an environment, ordinary objects and happenings cause sounds.

Recall that what you perceive as the duration of a sound is in fact the duration of the process of sound wave production.

Since the event in which sound waves are produced occupies a role causally intermediate to ordinary sound-producing events, such as collisions and strummings, and subsequent sound waves throughout the medium, this event plays a centrally important part in developing the theory of sounds. My claim is that such events are strong candidates for the particular events that are the sounds.

Consider the tuning fork. The sound, I propose, is the event of the tuning fork's disturbing the medium. According to this way of articulating the proposal that sounds are events, particular sounds are events of oscillating or interacting bodies disturbing or setting a surrounding medium into wave motion. This event occupies the appropriate causally intermediate role between the everyday events that cause the sounds and the compression waves that travel through the medium bearing the marks of sounds and producing experiences. If a sound just is an object or interacting objects disturbing the surrounding medium in a wavelike or periodic manner, then sounds do not travel through the medium but remain stationary relative to their sources. A sound unfolds over time at a location determined by the sound source. It does not travel through the medium, but it necessarily involves a medium. If sounds are the immediate objects of hearing, such disturbing events are the best candidates for the sounds. Its creating the disturbance constitutes the tuning fork's sounding.

The view that sounds are events of objects or interacting bodies disturbing a surrounding medium in wavelike fashion is not without historical precedent. The common interpretation of Aristotle is that he held a version of the received view according to which sounds are waves (see Pasnau 1999, 2000). *De Anima* says that 'sound is a particular movement of air' (II, 8, 420b10). This hints at a wave conception of sounds. Aristotle, however, understands 'movement' as active, which yields a view that differs in important respects from the view that the sound is the motion of the air. The idea of this alternative event account is to treat 'movement' as the nominalization of a transitive verb and to focus on constructions like '*x* moves *y*' instead of '*y* is moving'. The sound on this understanding is the movement of the air by the activities of ordinary bodies: 'For sound is the movement of that which can be moved in the way in which things rebound from smooth surfaces when someone strikes them' (420b20). We can understand 'disturbance' in a similar way as what takes place when *x* disturbs *y*, as the disturbance of *y* by *x*, rather than as the result of that event of interaction. The sound is the air's being disturbed by the motion of an object. A sound is not motion in the medium, but the activity of one thing's moving or disturbing another. This is not the wave account that most attribute to Aristotle, but the beginnings of the event theory of sound.

According to this account, sounds are particular events of a certain kind. They are events in which a moving object disturbs a surrounding medium and sets it moving. The strikings and crashings are not the sounds, but are the causes of sounds. The waves in the medium are not the sounds themselves, but are the effects of sounds. Sounds so conceived possess the properties we hear sounds as possessing: pitch, timbre, loudness, duration, and spatial location. When all goes well in ordinary auditory perception, we hear sounds as they are.

This characterization of sounds as events of objects' disturbing a medium amounts to a sketch of what sounds are. The rest of this book is an attempt to articulate and defend this conception as the most promising account of the metaphysics and perception of sounds. To begin to answer more fully the question 'What is a sound?' requires a more careful consideration of the individuation conditions of sounds understood as events.

5.3 Individuating Sounds

The event understanding counts among its greatest strengths the resources to capture convincingly the conditions under which sounds are identified and individuated from each other. I have mentioned already that sounds have durations, are capable of surviving changes to their properties and qualities across time, have stable distal locations, require a medium, and occupy distinctive causal roles. The disturbance event account is tailored to meet these criteria.

This event account, furthermore, individuates sounds primarily in terms of their causal sources and their spatio-temporal boundaries. A medium-disturbing by an item or body requires for its identity a given instance in which that item or body interacts with the medium. At a given time, therefore, sounds are distinct if their causal sources differ. More care is needed with the other direction of the conditional. When we understand the causal source as the very specific non-sound event that brings about the sound—the precise cause of the sounding, such as one of the fork's tines oscillating during an interval or the scraping of a single fingernail on a chalkboard—sounds at a time are distinct only if their causal sources differ. Nothing, on the other hand, prevents objects or coarsely specified events, such as automobiles or sporting events, from having multiple sounds at a time. Since sounds occur where things interact with the medium, the causal source constraint on sound identity at a time intuitively means that a sound must also be spatially continuous. Temporal discontinuities, in addition, mark distinct soundings. In order to count as the very same sound particular through time, therefore, a candidate requires the very same causal source, and must be spatially and temporally continuous throughout its entire history. A change in causal source, or a spatial or temporal discontinuity is sufficient for there to be discrete sound particulars.

Qualitative resemblance, however, is neither necessary nor sufficient for numerical identity of a sound. A temporally seamless transition from one instrument's playing a C sharp to another instrument's playing a C sharp involves numerically distinct sounds of the same sound type. Numerically distinct instances of qualitatively similar sounds are *the same sound* in nothing stronger than a qualitative sense. Temporally discontinuous soundings from the same source likewise are at most qualitatively identical. But when a single instrument seamlessly shifts from playing C sharp to playing B, only its state of sounding changes. There is still a single sound event of which each note instance is a part, and so each note instance is part of a single continuous sound. A sound can extend over considerable time and might change a great deal qualitatively. It may at times be loud and low-pitched; at times it may be soft and high-pitched. As long as the causal source remains the same and the sound is spatially and temporally continuous, it remains a single sound. Difficult cases for the spatial and temporal criteria, such as a tele-transported or time-traveling trumpeter, may of course arise. These cases should be decided by appeal to whether the causal source criterion is satisfied. When the causal source is numerically identical, spatial and temporal continuity from the point of view of the source may obtain and resolve the question in favor of identity. Such identity and individuation criteria stem from the account of sounds as disturbance events.

None of these considerations rules out that there might be complex sounds comprised of distinct sounds from a number of sources arranged either across space, over time, or both, as when an orchestra plays. Complex sounds might even include periods of silence; consider the sound of a song or of a spoken sentence. Complex sounds, however, are complex events constituted by many distinct sounds. There may be many different justifiable ways of counting sounds in these kinds of cases, but ways of counting complex sounds are intelligible because they invoke complex event types, such as that of a bus engine revving, or complex sound universals, such as those involved in musical compositions. The ways of counting or individuating sounds may differ depending on one's purpose. Understanding the metaphysics of music or of human speech sounds differs from developing an account of dolphin sounds because the kinds and complexity of the sound events of interest to each enterprise differ.

What is striking, however, is that disputes over individuation principles for sounds across these scenarios and disagreements about the number of sounds one has heard, mirror disputes about individuating or counting events themselves—disputes that are infamously difficult to resolve. This makes it a virtue of the event model of sounds that it leaves room for disputes about how many sounds have occurred, since it inherits that feature from questions and uncertainty about event counting and individuation. We are on just the right track, this suggests, when we construe sounds as events of a certain sort.

5.4 Two Objections

I have proposed that sounds are particular events in which a medium is disturbed or set into motion in a wavelike manner by the activities of objects or interacting bodies. Two initial objections concern features central to this account of the nature of sounds and of our capacity to perceive them. These objections must be addressed before we move on to develop the view and to consider further objections that concern the account's explanatory resources.

Are Disturbances Events?

One might worry that there are no such events. There is the event of the object or objects vibrating and there is the event of wave motion in the medium. The vibration causes the waves, the objection goes, but there is no additional event that counts as the disturbing of the medium by the vibrating object. Such an event would be the event of the vibration causing the waves. The worry is a fear of regress. If there is an event of the oscillations of the body causing the waves, what is the cause of this third, relational event? If it is the first, is there a further fourth event of the first causing the third? And so on.

But the disturbance events I have described are no more objectionable from a theoretical standpoint than the collision of two billiard balls, the opening of a drawer, the impact of a diver upon water, or the reflection of light from a surface. Medium-disturbing events are a kind of collision or interaction between objects and a medium. If there are such things as collisions or pushings, and if they are events, then the disturbance events I identify with the sounds are unproblematic. They have a natural description: they are events in which mechanical energy is transferred from one medium into another. They occur in a particular region of space and time; they involve the instantiation of a relational property; they have distinct causes and effects. Each theory of events may differ in its precise characterization of disturbance events, but to deny that there are such events as collisions or disturbings is to fall prey to the demands of an overly sparse ontology at the expense of common sense. Here are three ways to respond to the worry short of eliminating most of the interesting events.

First response The disturbing event is a causal event. It is identical with the event of a vibrating object causing the event of a medium exhibiting wave motion. But the object's vibrating does not cause the disturbing, so there is no regress. Rather, the disturbing is a complex event comprised of other events and that shares a cause with the object's vibrating. Ordinary sound generating events like strummings and breakings cause both the vibration events and the disturbing events the vibrations partly comprise. Problem: Sounds then do not cause waves, since the wave event partly comprises the disturbing. Solution: The disturbing event requires only a minimal wave event as participant, not the entire event of a wave propagating through the medium, which continues to occur long after the disturbing has ended. Though our talk may leave indeterminate what counts as a minimal participant, it differs from the complete wave event, the location of which changes over time. This leaves room for the disturbance event to cause subsequent stages of the wave event in the medium, which occur at locations successively distant from the source.

Second response The disturbing event is a relational event that involves two particulars: the object and the medium. But the disturbing is not identical with the vibration event's causing the wave event in the medium. The disturbing event is not constituted by those two events, but it shares constituents—the object and the medium—with those events. It differs from the causing because it involves the instantiation of a distinct relational property by the object and medium, the property they instantiate just in case they interact and that interaction has wavelike characteristics such as periodicity and amplitude. Several options exist for avoiding the regress. Perhaps no regress threatens because the structure of causal explanation differs among these different sorts of events. An object's vibrations might simply not be among the causes of its interactions. Perhaps, on the other hand, the disturbance event is causally intermediate to the vibrating and the waves, but there simply is no further event intermediate between the vibrating and the disturbing. Perhaps, given the increasingly fine grain of scientific explanations, there is a further event at an even finer grain of description to be revealed. That there are increasingly precise characterizations of events in the causal sequence is a matter of the structure of processes in nature that constitute observable events. The regress is only apparent. Its appearance results from this structure.

Third response The disturbing is identical with the vibration of the object in the presence of a medium. Vibrating is not disturbing since the former, but not the latter, could occur in the absence of a medium. Vibrating, and vibrating in the presence of a surrounding medium, differ in attributes including audible qualities. Vibrating in the presence of a medium, however, is disturbing, and the two share common causes. The vibrating does not cause the disturbing, so there is no regress, but the disturbing does cause the waves.

Can Disturbances Be Perceived?

A related worry is whether we can perceive such events as collisions or disturbings. This objection may concede that there are events of collision or interaction, but resists admitting that they enter the contents of perceptual experience. The experience, it may be argued, is neutral on the question whether interaction between the objects and a medium takes place. Since auditory experience is not neutral on whether a sound takes place, sounds are not disturbance events.

In response, even if the object of the experience of a sound is a disturbance event, one need not experience it *as* a disturbance event or as an interaction. Though in a sense accessible to the subject the experience is neutral on whether a disturbance event has taken place, the experience is in another sense not neutral. That is, the experience may not present the sound in a way that one might recognize it to be a disturbance event just by perceiving it, though the sound perceived is in fact a disturbance event. One recognizes a sound by its audible attributes such as pitch, timbre, location, and patterns of change to these attributes through time. Such qualities need not reveal a medium-disturbing event as an interaction between an object and a medium.

The objection, however, might be reformulated. Disturbance events, one might contend, are not, strictly speaking, among the sorts of things that could be objects of perception. All we perceive, for example, is the motion of one billiard ball followed by the motion of another. One simply cannot perceive such things as interactions or disturbings. At most, one perceives the relata of interactions; interactions themselves have no perceptual significance. If sounds were medium-disturbing events, we would not be able to perceive them. Since sounds have perceptual significance, sounds are not disturbance events.

Empirical evidence, however, strongly supports what is intuitively obvious. Interaction or its absence makes all the difference from the standpoint of perceptual experience. Brian Scholl, following Michotte (1963), details research that appears to demonstrate perceptual effects of perceived causality (Scholl and Tremoulet 2000; Scholl and Nakayama 2004). In one setup, subjects misperceive the distance between two converging objects when they seem to be interacting.

> Here we report a novel illusion, wherein the perception of causality affects the perceived spatial relations among two objects involved in a collision event: observers systematically underestimate the amount of overlap between two items in an event which is seen as a causal collision. (Scholl and Nakayama 2004: 455)

In addition, Bertelson and de Gelder (2004: 147) describe the following surprising perceptual phenomenon. Two simple visual objects that approach each other, coincide, and continue to move along a straight path seem visually to pass through one another when viewed alone. However, when either a sound or a flash accompanies the coincidence of the two simple visual objects, in many cases (60 per cent) the objects are seen to collide and rebound from each other (Sekuler et al. 1997; Watanabe and Shimojo 2000). These sorts of results appear to show that interaction or causation plays a significant role at the perceptual level, and not merely in subsequent inference or cognitive processing. Interaction impacts perception and experience. 'Phenomenology also supports this view: like the perception of faces or words, the perception of causality in collision events seems largely instantaneous, automatic, and irresistible' (Scholl and Nakayama 2004: 455). I would go one step further. It is the perception of the collision event, and the perception of the collision event as causal that characterizes the distinctive phenomenology of perceived interactions.[1]

Still, since in each of these cases we perceive the relata of the interaction, one might maintain that to perceive an interaction—for that interaction to count among the things one perceptually experiences—requires perceiving its relata. While this might be the norm in the case of visible objects, nothing prevents interaction itself from having perceptual significance in absence of perceptual awareness of what interacts. One might just

[1] Susanna Siegel (2005, 2008) also has recently argued on phenomenological grounds that visual experience represents causation.

be aware of a flash, an explosion, or a collision simply as an event or occurrence without experiencing what is flashing, exploding, or colliding. Likewise, one might be aware of a disturbance event through experiencing its distinctive audible qualities, while immediately experiencing neither the object nor the medium. Here, in particular, abandoning presumptions grounded in the visual experience of commonplace objects bears fruit and reveals a novel dimension of perceptual experience.

5.5 Sounds, Waves, and Experience

I have proposed a theory according to which sounds are audible events that occur just when objects disturb or interact in wavelike fashion with a surrounding medium. According to this account, we hear sounds because the auditory system detects or responds to waves. Though I have given both a priori and a posteriori arguments against identifying sounds with pressure waves that travel through a medium, does this account of sound perception nonetheless imply that we hear waves? In my view, it does not. Sounds are, in the sense discussed earlier, the immediate objects of auditory experiences. This does not rule out that we might hear things or events other than sounds. It does, however, mean that whatever we hear that is not a sound, such as a tuba or a brawl, we hear it in virtue of hearing a sound. Since sounds are not identical with waves, we do not immediately perceive waves. Since waves themselves are not among the things we hear in virtue of hearing a sound, we do not hear pressure waves.

According to this account, we may say that the event I identify as the sound is discerned by way of its audible qualities. Still, one might wonder whether I have characterized the disturbance event in a way that makes sound waves constitutive of the sound event. The waves, however, need not constitute that event, since the waves are causal by-products of the interaction between the bodies and the medium. The sound plays a causally intermediate role between ordinary events like breakings or crashes and the event of waves propagating through the medium. What is constitutive of the sound event is the wavelike or periodic activity that occurs at the surface of interaction between the vibrating object and the medium. Strictly speaking, the disturbing event is a relational event that involves the object and the medium. But it need not be constituted by the wave that subsequently speeds through the medium. It is important to distinguish the wavelike or periodic character of this movement from the resulting wave. It is one thing to say that a medium is a necessary condition for the existence of a sound, since the sound constitutively requires the interaction between a body and a medium, and quite another to say that a wave constitutes the sound.

I have argued up to this point that sounds are non-mental particulars, and that they are events that take time, occur at distal locations, and involve a medium. The event conception of sounds according to which sounds are medium-disturbing events is most naturally suited to explain what is most distinctive about sounds and the experience of sounds. It captures the decidedly temporal characteristics not just of auditory experiences, but of sounds as we perceptually experience them to be. It provides a phenomenologically plausible account of how we learn the locations of things and happenings in the environment through audition. It secures the medium-relative nature of audible qualities.

Satisfying these constraints, however, does not suffice for success in the theory of sounds. A realist theory of sounds must convincingly account for a host of sound-related phenomena, including the audible qualities, echoes, interference, transmission through barriers, and Doppler effects, which present obstacles to any theory that situates sounds in the world independent of the subjects of auditory experiences.

In the chapters that follow, it will become evident that the event account proves to be an illuminating and explanatorily resourceful account of sounds and sound-related phenomena. The framework that results from understanding sounds as distally located events captures perceptually salient aspects of the world of sounds and makes clear the informational value of audition in complex environments. The event understanding of sounds furnishes a principled account of the circumstances in which we perceive with success and in which we fall prey to auditory illusions; it surpasses even the most service-worn alternatives.

6

Audible Qualities

Particular sounds have pitch, timbre, and loudness. A tuba's notes are lower pitched than a flute's, though the fuzz from an untuned radio has no discernible pitch. Middle C played on a piano has a character or timbre that differs from middle C played on a trumpet. Jet planes make louder sounds than tractors. So it seems in audition.

Thus far I have discussed the nature of sounds themselves. Philosophical tradition, I have claimed, groups the sounds, along with colors, tastes, and smells, among the sensible qualities. I have argued instead that sounds are particular events of a certain sort. Sounds are events in which objects or interacting bodies disturb a surrounding medium in a wavelike manner. Such events count among their effects longitudinal pressure waves that propagate through the medium.

Though I have argued that sounds are not qualities or properties at all, sounds do bear audible attributes whose metaphysical footprints more closely resemble those of the visible or olfactory qualities. A particular individual might change from being scarlet or sweet at one time to being indigo or sour at another, and a sound might change from being high-pitched and loud to being low-pitched and quiet. Distinct surfaces share hue or lightness, and distinct sounds share pitch or loudness. Just as the colors cannot be tasted and the tastes cannot be seen, pitches cannot be smelled and timbres cannot be seen. Perceivers exhibit subtle differences in judging visually what counts as unique green, and listeners differ in what sounds count audibly as middle C.

Just as cilantro tastes soapy to some tasters, teenagers hear certain tones that adults do not—enterprising youths recently have deployed a cellular phone ring tone that capitalizes on the fact that, because of a form of age-related hearing loss called 'presbycusis', tones at around 17 kilohertz are inaudible to most adults (Vitello 2006).

This suggests that accounts of secondary or sensible qualities familiar from debates over the nature of color should furnish a battery of arguments and theories concerning the natures of audible qualities. Audible qualities, following such accounts, might be dispositions of sounds to produce specific sorts of auditory experiences; objective physical attributes of sounds; primitive, simple, unanalyzable, or manifest properties of sounds; or even purely subjective, sensory, or psychological features. Since sounds have among their apparent audible attributes pitch, timbre, loudness, and the like, the theory of sounds must accommodate or explain away the audible qualities. What exactly are pitch, timbre, and loudness? That is, what are the natures of the audible qualities?

Physics and psychophysics have taught us that, given how audition works, properties such as (roughly, for now) frequency, wave shape, and intensity causally determine which pitch, timbre, and loudness a sound auditorily appears to have. In light of this, we might ask exactly how the audible qualities are related to the physical properties of frequency, wave shape, and intensity. Because physical properties such as frequency, wave shape, and intensity causally explain auditory experiences of pitch, timbre, and loudness, providing a theory of the audible qualities we perceive is tantamount to explaining how pitch, timbre, and loudness themselves depend upon such physical characteristics. If, for instance, pitches are dispositions to produce experiences as of pitch in certain sorts of perceivers, then such dispositions have frequencies as their categorical bases or grounds because frequency is what disposes bearers of audible qualities to stimulate pitch experiences. Frequency is the property of sounds responsible for manifestations of the disposition.[1] If pitches just are the physical properties causally responsible for pitch experiences, then pitches are identical to frequencies. If pitches are simple or primitive properties of sounds, then pitches nonetheless supervene upon, are constituted by, or correlate universally with frequencies (if pitches ever are instantiated). If apparent pitches are projected, purely sensory features, then frequency is the property that stimulates the distinctive sort of auditory experience. Whether audible qualities are dispositions, physical properties, primitive qualities, or projected sensory qualities, their instances depend upon the physical properties that explain the occurrence of audible quality experiences. Whether the relationship is grounding, identity, universal correlation, or mere causation, an account of audible qualities turns upon the kind of role it ascribes to physical attributes of sounds in determining when an audible quality is instantiated.

Theories of audible qualities that do not impute to audible quality experience a massive projective illusion—those that situate audible qualities outside the minds of hearing subjects, either as dispositions, physical properties, or simple properties—imply that which audible qualities are instantiated in the world of sounds depends, in a sense stronger than causation, upon which physical attributes are instantiated. The sense of dependence is that indiscernibility with respect to the relevant physical attributes instantiated in a region guarantees indiscernibility with respect to audible qualities instantiated in that region. For instance, no difference in pitch occurs without a difference in frequency; no difference in timbre occurs without a difference in wave shape and spectral composition; no difference in loudness occurs without a difference in intensity. That is, according to theories that do not posit radical error or eliminate audible qualities, audible qualities supervene upon physical attributes such as frequency, wave shape, and intensity. Particular

[1] Again, this is oversimplified for now. See below for further development concerning the relationship between pitch experiences and frequencies.

accounts of audible qualities differ in the modal strength of the guarantee.[2]

I do not wish to replicate the philosophical debate concerning the nature of color in every detail.[3] The event theory I have advanced should be flexible. It should not commit one to any particular theory of audible qualities, any more than those who would identify sounds with waves are so committed.

The account of sounds as medium-disturbing events therefore faces a different, pressing issue. Unless audible quality experience is a 'perfectly monstrous illusion' (Byrne 2003), instances of the audible qualities of pitch, timbre, and loudness supervene upon instances of the physical characteristics of frequency, wave shape, and intensity. Frequency, wave shape, and intensity, however, are commonly understood as attributes of waves. The audible qualities depend upon properties ordinarily ascribed to sound *waves*, and whether they are instantiated in a region is a matter of whether the corresponding physical attributes are instantiated in that region. The view of sounds I have advanced, however, holds that sounds are events that occur when objects and bodies interact with a surrounding medium. Sounds, according to this theory, are not identical with and do not supervene upon sound waves in the medium. Does it follow that sounds as I have characterized them lack audible qualities? If pitch, timbre, and loudness are properties of sounds, they must be properties of the events I have identified as sounds. Because the audible qualities depend upon frequency, spectral composition, and intensity, the success of this proposal

[2] In what sense do all accounts agree that audible qualities supervene upon physical properties? A sense that, as we should expect, is very weak, indeed. A form of primitivism might simply hold that the audible qualities and physical properties are nomologically coextensive.

[3] See O'Callaghan (2002) and work in preparation for a full treatment of this issue. I argued in (2002) that audible qualities are complex physical features of sounds, responded to objections based on psychophysical data that purport to show that pitch does not correspond to frequency, and developed the view to address philosophical objections, including those based on an auditory form of spectral shift that occurs in cochlear implant recipients. In work in preparation, I propose a different, simple understanding of pitch as periodicity and claim that it handles most of the standard arguments against color objectivism more easily than do extant forms of color physicalism.

relies on there being a well-motivated ascription of frequency, spectral composition, and intensity to sound events as I have characterized them. Otherwise, the event understanding entails an unintended error theory.

In what follows, I want to show why the dependence of audible qualities upon physical characteristics needs to be preserved, and how to preserve it if sounds are distal events. Once we know how to ascribe the desired attributes to sounds construed as distal disturbance events, room exists for a variety of theories of pitch, timbre, and loudness. More important, we shall discover that accounts that locate the audible qualities and their bases at distal locations near the sources of sounds surpass wave-based accounts in explaining a number of striking perceptual phenomena, including constancy effects for loudness, timbre, and pitch. The event understanding of sounds therefore delivers a better account of the bases of audible qualities. The explanatory benefit derives from dissociating changes to the audible qualities from changes that occur entirely in the medium and tying them instead to changes to the medium-disturbing events.

6.1 Periodicity and Pitch

A very simple *frequency theory* of pitch can be extracted from the story we learned about sounds in primary school. We were taught that the pitch of a sound is the frequency of a pressure wave that travels through a medium such as air, water, or helium, where frequency is just the number of cycles of motion per second, measured in *hertz*. Whatever doubts we have about the natures of colors, accepting the frequency theory makes us happy to say that pitches reside in the world beyond our minds. What we discern when we hear pitch just is the property of having some frequency. When we tune a trombone, one way to get its pitch right is by adjusting the frequency of an F note with an electronic tuner.

This frequency theory accounts for salient aspects of the experience of pitch. It explains the linear ordering of pitches: as frequency increases, so does perceived pitch. A natural account of the musical intervals or relations, which are pitch relations, also follows from the frequency theory of pitch. Small whole number frequency ratios form the bases of the octave (1:2), fifth (2:3), fourth (3:4), and so on. One gets the palpable sense that the natures of such audible relations are revealed by this discovery. The linear ordering of pitches can be adapted to incorporate the musical relations. Imagine twisting the line into a helix around a vertical axis so that tones separated by an octave—tones that are the same in the relevant respect—have the same angle of rotation about the vertical axis. So, for example, each C falls on a vertical line in ascending order of height. According to this representation of pitch, the same notes at different pitch heights have the same angular position, and each rotation of the helix corresponds to an octave.

The frequency theory, nonetheless, is far too simple. The identification of pitch with frequency approaches adequacy for *pure* or *sinusoidal* tones—sounds whose accompanying waves are constituted by sinusoidal pressure variations, as in Figure 5. For sinusoidal tones, increasing frequency increases perceived pitch, and decreasing frequency decreases perceived pitch.

Pure sinusoidal tones, however, rarely are encountered in nature, and many complex sounds that are not themselves sinusoids are perceived to have pitch. Fourier showed that any complex sound

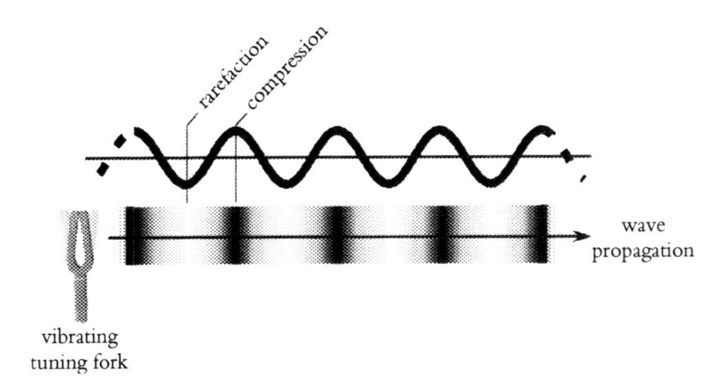

Figure 5. Sinusoidal motion

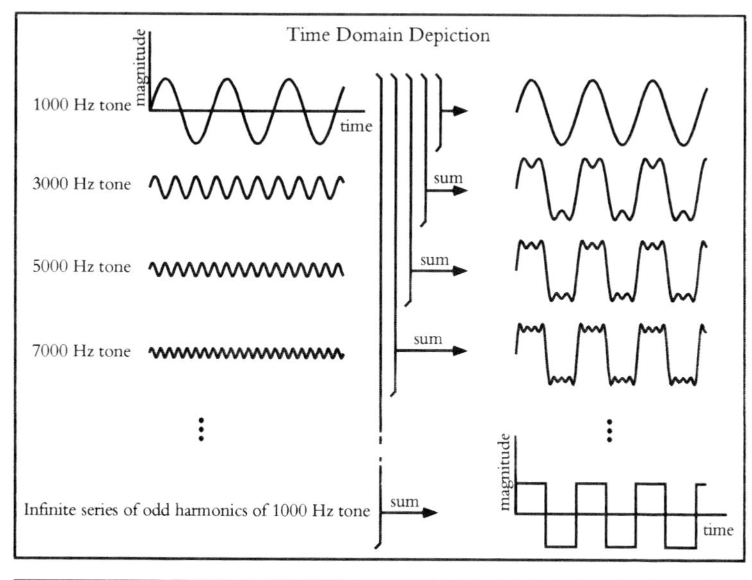

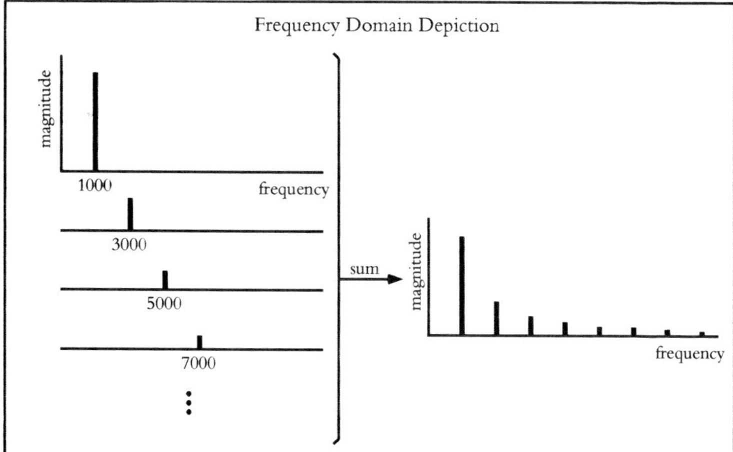

Figure 6. Fourier composition

which is not itself a sinusoid can be analyzed in terms of sinusoids of various frequencies in differing proportions. Figure 6 illustrates this for a complex square wave. Fourier's theorem applies in virtue of the additive principles of wavelike motion. So we can characterize any sound in terms of a *spectral profile* by citing the amplitude or intensity of each of its sinusoidal constituents.[4]

The constitutive role of individual sinusoids can be taken quite literally: the phenomenon of *sympathetic resonance* demonstrates that complex tones are made up of sinusoids of different frequencies. A pure sinusoidal tone from one tuning fork causes a second tuning fork with the same characteristic frequency to begin sounding in virtue of its own induced sinusoidal motion. A complex tone from a human voice, one of whose Fourier-predicted components shares a tuning fork's characteristic frequency, induces that tuning fork to resonate just as a sinusoidal tone of that frequency does. But a complex sound without a Fourier component at the tuning fork's characteristic frequency induces at best a weakened resonant sounding. The best explanation of sympathetic resonance is that a tuning fork's resonant sounding is caused by a sinusoidal constituent which is genuinely present in the complex sound.

Pure sinusoidal tones are a variety of *periodic* sound. That is, they repeat their simple motion at regular intervals. Some complex sounds, however, also are periodic and repeat a certain pattern, which might be very complex, at some interval. If a complex signal is periodic and repeats at regular intervals, the frequency of each of its components (or *partials*) must have an integer-multiple relationship to a *fundamental* frequency (frequency is period's inverse). This fact accounts for the repetition of the pattern. The fundamental frequency of a complex sound is the greatest common whole number factor of the sound's constitutive frequencies. Thus, the fundamental frequency of the complex (square) tone in Figure 6 is 1000 hertz. Periodic complex sounds are called 'harmonic', and each of the upper partials is an upper harmonic.

[4] Strictly speaking, we cite the maximum or *root–mean–square* amplitude or intensity. The root-mean-square amplitude is a measure of the average magnitude of a varying instantaneous amplitude. It is given by the square root of the mean of the squared instantaneous amplitudes. For a sinusoid, the root-mean-square value is 0.707 times the peak amplitude.

As a matter of empirical fact, just the periodic sounds have pitch. Though periodic tones may occur within more 'messy' or 'noisy' sources and thus cause pitch experiences, and sounds that are approximately but not exactly periodic may on occasion appear to have pitch, pitched tones generally are periodic sounds. The class of sounds with pitch is the class of periodic sounds.[5]

The perceived pitch of a periodic complex sound furthermore matches that of a simple sinusoidal tone at the fundamental frequency. Likewise, complex periodic sounds that share fundamental frequency match in pitch even if they share no spectral constituents. For example, a 100-hertz sinusoid, a complex sound with constituents at 200 and 300 hertz, and a complex sound with constituents at 1000 and 1100 hertz all match in pitch. So the case of pitch presents something like that of metamerism for colors. Just as surfaces whose spectral reflectance profiles differ greatly may nonetheless match in color, sounds whose spectral profiles differ greatly may match in pitch. Individual pitches, like colors, might therefore be understood in terms of spectral profile types (compare Byrne and Hilbert 1997*b*, 2003; Tye 2000; Bradley and Tye 2001).

The difference between the cases of color and pitch, however, is that in the case of pitch there exists a straightforward unifying characteristic independent of pitch experiences that accounts for membership in the type. Members of a given spectral profile type whose members match in pitch share fundamental frequency or periodicity. Metamers present such a problem in the case of colors because it is difficult to motivate spectral reflectance profile types as natural, perceiver-independent properties. Such types seem arbitrary, gerrymandered, or unnatural apart from their significance to color perceivers. No such problem exists for pitches. A given pitch includes all sounds, simple and complex, that share a fundamental frequency. Such sounds may differ radically from each other in their constituent frequencies (and thus in timbre),

[5] Or a subset of the periodic sounds, since some periodic sounds are outside the range of hearing. I am happy to admit pitched sounds no human hears.

but each has harmonics that are integer multiples of some common frequency, and each repeats at the same regular interval. What sounds that share pitch have in common is the repetition of some pattern, not necessarily the same pattern, at the same regular interval.

Pitch therefore depends upon periodicity. For a sound to have pitch requires that it be periodic in nature. A simple *periodicity theory* of pitch holds that the pitch of a sound is its periodicity, which is characterized by its fundamental frequency. A periodicity theory might also hold that some dependence relation weaker than identity holds. In either case, the pitch of a periodic sound is determined by the greatest whole number frequency by which the frequency of each of its sinusoidal components is divisible without remainder.

According to this formulation a sinusoidal component at the fundamental frequency may or *may not* actually be present in the complex sound. The phenomenon of the 'missing fundamental' demonstrates that the fundamental frequency component need not be present for a sound to have the same pitch as a sinusoid of that frequency. A tone with constituents at 1200, 1300, and 1400 hertz has the same perceptible pitch as a 100-hertz sinusoid (see Helmholtz 1877/1954, Schouten 1940). Telephones, which filter low frequencies, illustrate the principle. One hears a man's voice to have the same pitch in person and over the telephone, though the fundamental is absent from the telephone loudspeaker's sound.[6]

[6] Terhardt (1974, 1979) has argued that this simple conception is inadequate to account for all types of pitch phenomena. He has distinguished between *spectral* pitch, which is pitch heard in virtue of constituent frequencies actually present in the sound, and *virtual* pitch, which is heard despite the absence of a spectral constituent. Terhardt has pointed out that a view analogous to the simple periodicity theory fails to account for the fact that a single complex sound is often heard to have multiple pitches, determined by both spectral and virtual pitches. He also has noted that *both* types of pitch may be heard simultaneously, even at the same frequency. Terhardt's conception depends, however, upon individual pitches, spectral and virtual, being determined by constitutive frequencies present in the sound. Indeed, individual spectral and virtual pitches are identified with particular frequencies on his account. So the simple theory that pitch depends on a sound's periodicity may not accommodate all salient pitch phenomena, but it is sufficient for our purposes that pitches depend upon particular frequencies, and that pitch determination (even if a single sound has

This fact led Helmholtz and others to believe mistakenly that the pitch experienced at the missing fundamental, or 'residue pitch' (in contrast to the 'spectral pitch' that corresponds to a sinusoidal component present in the sound), was illusory pitch. The missing fundamental introduces difficult complications from the standpoint of the mechanisms of pitch perception, but it is no more an illusion if pitch depends upon periodicity than is hearing the pitch of a sinusoid.

The periodicity theory grounds pitch in the periodic character of a sound, which is determined by its fundamental frequency and depends upon the frequencies of a sound's sinusoidal constituents. It follows that pure sinusoidal sounds and complex sounds such as those that exhibit sawtooth or square wave patterns can share the same pitch in virtue of sharing a fundamental frequency.

The periodicity theory has a number of advantages as an account of pitch. It counts pitches among the attributes of sounds that do not depend on the auditory experiences of subjects, but instead are among the potential causes of pitch experiences. Since the periodicity theory bases determinate pitches on particular fundamental frequencies, it preserves the frequency theory's influential account of the musical relations as grounded in small whole number frequency ratios.

It has the further advantage of being supported by the physiology of auditory sensation. The basilar membrane, located within the cochlea of the inner ear, is like a long trapezoidal ribbon, different parts of which most actively resonate in sympathy with a particular sinusoidal frequency. The basilar membrane thus performs a sort of Fourier analysis, decomposing a complex signal into its sinusoidal components. This spectral information is converted into electrical potentials by the hair cells, which activate auditory neurons. Given frequencies activate portions of the auditory

both spectral and virtual pitches) is a matter that depends entirely upon which frequencies are present in a complex sound.

nerve 'tuned' to those frequencies, and this tuned or *tonotopic* organization continues up through the auditory cortex where higher cognitive processes determine pitch or fundamental frequency from spectral information about the sound. 'Every major nucleus between the cochlea and the cortex has been found to be cochleotopically [tonotopically] organized' (A. Palmer 1995: 81). Even a cat's cortex is arranged in individual columns which are tuned to characteristic frequencies (Woolsey 1960). The significance of Fourier decomposition at the basilar membrane and the tonotopic organization of the auditory system is that frequency parameters appear to be preserved and represented through various stages of auditory processing. In fact, electrodes can recover and reproduce a sound presented to the ear from the subsequent auditory nerve signals. This is the so-called Wever-Bray effect.

> Wever and Bray reported that if the electrical activity picked up from the cat's auditory nerve is amplified and redirected to a loudspeaker, then one can talk into the animal's ear and simultaneously hear himself over the speaker. (Gelfand 1998: 136)

It appears that pitch perception depends upon determining the frequencies present in a complex signal. If pitch depends upon periodicity, the problem of pitch perception is physiologically tractable.

6.2 Pitch and the Event Theory of Sounds

When discussing sounds and their pitch, frequency and thus periodicity are commonly understood as properties of sound waves. If pitch must be understood in terms of frequencies, and if frequency is a property of waves, then pitch locally supervenes upon a property of waves. But the view I have proposed holds that sounds are events that occur in locations that differ from those of sound waves. Unless sounds lack pitch, pitch must be a property of the events I have identified as sounds. I do not wish to commit myself just in virtue of my theory of sounds to an error theory of pitch. How can an account according to which sounds are distal events of objects' interacting with a surrounding medium preserve the dependence of pitch upon periodicity and frequency while ascribing pitch to sounds? The success of my proposal relies on there being a well-motivated ascription of frequency and periodicity to sound events as I have characterized them.

Consider frequency as it is ascribed to waves. Frequency is a function of positions and times. The positions of particles in the medium at different times characterize their vibratory motion, which in turn determines the pressure at a point in the medium, as in Figure 5. Frequency is the number of cycles of motion per unit of time that particles undergo, which we can track by noting their points of maximum displacement. But particle displacements are caused by the activities of the bodies that disturb them, and periodic sounding bodies themselves can be ascribed a frequency. Since the displacement of a vibrating body immediately causes surrounding particles to be displaced, as illustrated in Figure 7, the frequency with which a sounding body reaches one point of maximum displacement corresponds directly to the frequency of the resulting wave in the medium. The periodicity of the sounding body matches that of the resulting wave.

Many sounding bodies have complex vibration patterns with different modes of sinusoidal vibration that contribute to the overall shape of the resulting wave. In this case, a sounding can be ascribed a complex pattern of frequencies that determines its overall spectral composition. This, too, will match the vibration patterns of particles in the medium, and, thus, of the wave.

For simplicity, we can treat the location of the disturbing event at a time as the surface of interaction between the object and the

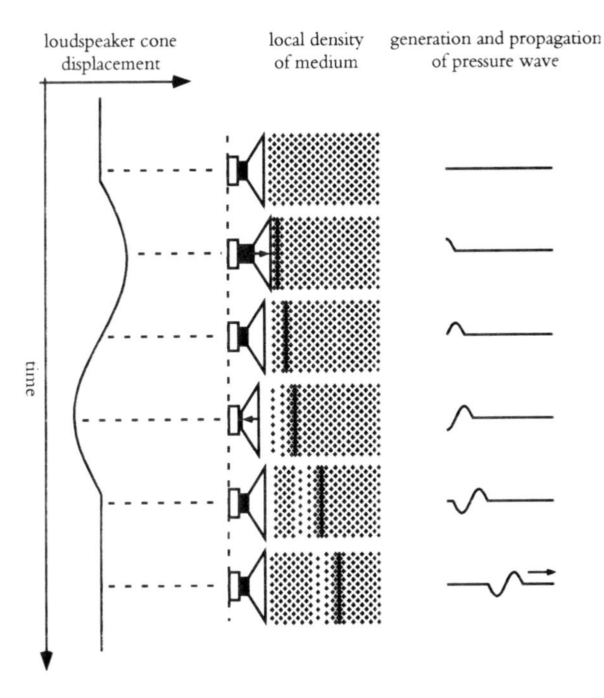

Figure 7. Vibration transmission and period matching

medium. Compare Jonathan Bennett's remarks about the locations of events:

> The 'location' of an event is its spatiotemporal location, i.e. where and when it occurs ... A zone may be sizeless along one or more of its dimensions: ... some fill spatial volumes and presumably others occupy only planes, lines, even points. (Bennett 1988: 12)

When tracked through time, this surface itself vibrates with a frequency that corresponds to the vibration frequency of particles that constitute the surrounding medium. Let the *frequency of the sound event* be defined as the vibration frequency of the disturbance surface. Since wavelength is inversely proportional to wave frequency (frequency (f) equals the speed of sound waves in the medium (v) divided by wavelength (λ), $f = v/\lambda$), the number of wave peaks that pass through a point in a medium during a given time interval just is the number created by the sounding body's maximum displacement during an earlier time interval of the same duration. Frequency thus is independent of the speed at which waves travel in a medium. The frequency of the sound event—that is, the frequency of the medium-disturbing event—is therefore identical to the frequency of the subsequent wave in ordinary circumstances. So the event view commonly ascribes to sounds frequencies that match those of their waves in the medium.[7]

Since pitch depends upon the periodicity of a sound, pitch depends on its frequency and spectral composition. The sound event's frequency and spectral composition, which are determined by its modes of vibration, ordinarily equal those of the resulting waves. Sound events exhibit periodicity and therefore possess pitch.

6.3 Loudness

Loudness is an attribute of sounds in virtue of which sources might seem more or less intense, as greater or softer in volume. Loudness depends upon the intensities of a sound's spectral constituents. Acoustical measurements of sound level at a given place in space frequently are expressed in decibels of intensity level (dB IL) (where I_0 is a reference or just-audible intensity level and I is root-mean-square intensity, dB IL $= 10 \log(I/I_0)$). Since intensity is proportional to the square of pressure, converting from decibels of intensity level to decibels of sound pressure level (dB SPL), another common measure of sound level, requires squaring pressure values. Intensity, a measure of energy flux, is defined as the power, or rate at which energy is emitted ($P = \Delta E/\Delta t$), passing through

[7] But see the discussion of Doppler effects in Chapter 7 for cases in which the event understanding correctly ascribes to sounds frequencies, and therefore pitches, that differ from those of the waves.

a surface per unit area ($I = P/A$). We can therefore express the intensity of a sound event as the power transmitted per unit area on the surface that gives the sound event's location. Physicists interested in sound production, in fact, often cite this quantity. Because the power transmitted through this surface depends upon the specific properties of the surrounding medium, the intensity of the sound event varies across different mediums. Since the loudness of a sound is a function of its energies or intensities at various frequencies, the loudness of a sound event is medium-dependent.

Sounds construed as medium-disturbing events therefore exhibit the patterns of activity that determine frequency and intensity, which in turn determine pitch and loudness. It follows on this account, however, that the effects of changes to sound waves as they travel through the medium do not alter the audible properties of a sound itself. Though wave distortions may affect the experience of pitch, the sound's pitch need not change. Though degradation of waves may affect the experience of loudness, the sound's loudness is unaffected by such changes. The event view thus delivers a more satisfactory account of a sound's loudness than does the wave view because the event view captures the phenomenon of *loudness constancy* (Zahorik and Wightman 2001). We do not perceive a sound itself to change in loudness when we move away from it—a speaker's voice does not itself get louder just because I move to the front of the room. I may be in a better position for optimal or comfortable hearing, but the qualities of the sound do not differ with distance from its source. According to the event view, a sound's intensity is a matter of the power through the surface that demarcates the disturbance event's location. And loudness is a complex function of the sound's intensity at various frequencies. But since intensity diminishes as the inverse square of distance from a sound source, the wave account, which attributes loudness to sound waves, does not capture loudness constancy. According to a wave-based theory of audible qualities, the loudness of a sound itself changes as the waves move away from their source. This fails to comport with the perception of loudness.

6.4 Timbre

Timbre has been dubbed 'the psychoacoustician's multidimensional wastebasket category' (McAdams and Bregman 1979). Fittingly, it has been timidly defined as that attribute in virtue of which sounds that share pitch and loudness might differ (ANSI 2004). Sometimes timbre is referred to generically as 'sound quality' or 'sound color'. However timbre is characterized, a tuba and a trumpet playing the same note at the same volume differ audibly. Two people singing that same note sound different from each other and from both the tuba and trumpet. Timbre is the audible difference. Timbre as such is critical to us when we identify the everyday objects and events that we are able to recognize just by hearing. Since the acoustical properties of audible objects and events depend upon characteristics of the interaction between participating objects and the surrounding medium, our ability to recognize sound sources thanks to hearing depends upon information about the characteristics of medium-disturbing events. Which characteristics?

Most prominent among features that contribute to sound and source recognition, and to timbre itself, is the spectral shape of the sound. Spectral shape is determined by the relative amplitude of each sinusoidal constituent across the spectrum. The vibration modes of a source determine a sound's spectral composition, and the tonotopic organization of the auditory system, which is required for pitch perception, preserves information about spectral constituents of a sound. Since medium-disturbing events have various modes of vibration at different amplitudes, spectral shape presents no unique problem.

Features also relevant to timbre and to source recognition include aspects of the onset (or attack) and offset (or decay) of particular sinusoidal constituents or frequency ranges; patterns of inharmonic noise, especially for natural events; transitions and overlap between sounds; and timing and rhythm of continuous, oscillating, or discrete sounds. Each aids subjects in some fashion to identify the sounds of particular musical instruments, speech, voices, natural objects, actions, and events (Handel 1995). Nevertheless, Handel claims, 'At this point, *no* known acoustic invariants can be said to underlie timbre' (Handel 1995: 441). The acoustic cues to timbre that ground identification of objects and events vary and depend on context: 'the duration, intensity, and frequency of the notes, the set of comparison sounds, the task, and the experience of the subjects all determine the outcomes' (441). Most striking, however, is that each cue depends upon the particular source's distinctive manner of disturbing the medium across a range of conditions. 'The cues that determine timbre quality are interdependent because all are determined by the method of sound production and the physical construction of the instrument' (441). Handel suggests that timbre is something like the distinctive *look* of a face. The look of a face is determined by the arrangement and characteristics of particular visible features. The timbre of a sound is determined by the arrangement and characteristics of audible features. The recognizable look of a face is constant across changes to profile and hairstyle, those due to age, and even mild plastic surgery. The timbre of a sound is constant across changes to frequency, intensity, attack, decay, and even mild vocal chord damage. This suggests that the timbre quality of a sound depends at least in part upon features of the source and the characteristic manner in which it disturbs the medium. That is, after all, what remains constant across changes to its determinate audible qualities. The uniformity of timbre across sounds and circumstances is best explained by constancy in factors beyond the attributes of waves. The event understanding according to which sounds are events of objects or interacting bodies disturbing a surrounding medium therefore better serves the needs of an account of timbre perception and recognition. Shifting focus from waves to activities of sound sources, and to the ways things interact with and disturb the surrounding medium, enables this event understanding of sounds to shed light upon regularities in auditory quality experience and thereby to capture critical aspects of what makes audition so informative. The event account has little to fear, and much to gain, in acknowledging the dependence of audible qualities upon physical attributes of sounds.

7

Sound-Related Phenomena: Transmission, Interference, and Doppler Effects

7.1 Explaining Sound-Related Phenomena

No theory that fails to address the facts about locational hearing can be a phenomenologically plausible account of the objects of auditory perception. Furthermore, no view that disregards the role of the medium in the generation of sounds can convincingly capture variation to the audible qualities of a sound across different mediums. The view according to which sounds are particular events of objects or interacting bodies disturbing a surrounding medium in a wavelike manner recognizes the distal locations of sounds while preserving the important truth in the wave conception that no sounds exist in vacuums. The event view thus avoids both horns of the dilemma over locating sounds in the sources or locating them in the medium.

The principal appeal of the view stems from its appreciation of the strikingly event-like characteristics of sounds. Sounds happen and take time to occur. Both the traditional visually influenced philosophical view of sounds as secondary qualities and the commonplace scientific view of sounds as waves that propagate through a medium fail because they do not, and perhaps cannot, fully confront the essential temporal characteristics of the immediate objects of auditory awareness. To be a sound as heard is to be a temporally extended particular whose identity is partly determined by a pattern of qualitative changes though time. A sound, moreover, is constitutively tied to the activities of and changes to objects and a medium. Without either sort of activity and change, there would be no sounds.

The discussion so far leaves unresolved a host of questions about pervasive sound-related phenomena. The familiar wave model is fantastically successful at explaining the audible effects of constructive and destructive interference; transmission through interfaces, such as between water and air, and through barriers, such as walls and windows; echoes and reflected sounds; and the Doppler effects. Each involves an auditory experience that may not comport with what occurs at the source. In short, local wave characteristics at the ears contribute to audible effects. Divorcing the sounds from the waves traveling through a medium means the distal event view (indeed, any view that locates sounds at or near their sources) owes equally explanatory accounts of these phenomena and of the related contents of perception.

The event understanding of sounds surpasses the wave view's success at convincingly accounting for such phenomena. The event view claims that sound waves transmit information about the sounds and their audible qualities. It therefore explains interference, transmission, Doppler effects, and echoes as stemming from wave phenomena that have nothing essentially to do with sounds themselves. This kind of explanation, I argue in this and subsequent chapters, succeeds where the wave conception of sounds falls short. The distal event theory not only delivers an account of why auditory experiences with particular characteristics arise as a result of conditions in the acoustic environment, but it also motivates a compelling understanding of the corresponding contents of auditory perception. It succeeds here precisely because it refrains from identifying the sounds with the waves. The event understanding not only allows for veridical auditory experience where the wave view does not—in the experience of location and duration—but predicts auditory illusions where the wave view cannot. It emerged in my discussion of the audible qualities of sounds that where wave-based accounts must attribute a veridical experience of wave characteristics, the event account may ascribe an illusory experience of a sound, where the illusion is due to wave effects or distortions. This, in certain circumstances, has benefits. The thesis I wish to develop by considering a number of cases is that the event theory of sounds furnishes a compelling and principled account of the conditions under which the auditory experience of a sound is veridical and of those under which it involves illusion.

The burden on the event theorist is not only to show that the preferred account is capable of explaining why a given experience arises in the face of some situation in the world of sounds, but also to motivate verdicts about the world of sounds in the face of sometimes confounding evidence from auditory experience. In arguing that the event view of sounds gracefully shoulders the explanatory burden, here and throughout I take as my foundation certain basic scientific explanations of the phenomena in question. The consequences of these scientific facts, however, are open to philosophical interpretation.

7.2 Transmission

So far, I have said that a sound is an event of a medium's being disturbed or set into motion in a particular way by the activities of an object, body, or mass. But this seems too lenient, to allow too many sounds. Consider the following two forms of objection.

Interface

Suppose you are underwater and hear the sound of something that happens in the air above, say, the striking of a bell. The event view seems to imply that there is a sound at the interface of the air and water since indeed there is a medium-affecting event there.

The air, a mass or body, sets the water, a medium, into motion. This is phenomenologically inaccurate. We do not hear the sound to be at the surface of the water; we hear it to be above in the air.[1]

Barrier

The preacher outside on Portland Street is loud. When I shut the window I do not hear him as well. The window muffles the sound. Nevertheless, the window also sets the medium inside the room into motion. According to the event view, is the sound located at the windowpane? We do not hear it as being there—the sound still seems outside.

In both forms, sound waves generated in one medium pass into another kind of medium. The first describes travel across a single interface; the second involves travel through a solid barrier. In each, at the relevant interface—the air–water interface in the first case and the window–room interface in the second—the motion of a body disturbs the medium it adjoins. Yet since we do not ordinarily take ourselves to hear sounds at such places, intuition has it that no sound occurs at either the interface or the barrier. Must the event theorist count these events as sounds?

In fact, there is a slightly more troubling extension of this objection. Sound waves are transmitted by means of collisions among the particles of a medium. If each one of these collisions is an event in which an object (a particle) disturbs a medium, then the event view seems to imply that each of *these* events is a sound. We might respond that the molecules do not count as *objects* or *bodies* in the ordinary sense enlisted in the conception of a sound. This is unsatisfactory. We are discussing ontology, and must not artificially limit the scope of what counts as an object or body in order to suit the purposes of our favorite theory. At any rate, this extension illustrates the potential extent of the difficulty that sound wave transmission poses for the event theorist.

[1] Here, and in the discussion that follows, I shall ignore misperception of the sound's location that results from refraction. This affects neither the argument nor the conclusion.

The problem of transmission is not unique to the event view. Each of the views canvassed faces a version of the objection. For instance, Pasnau's property view is in roughly the same straits as the event view. The property view implies that the sound is a property of the air mass in 'Interface' and of the windowpane in 'Barrier', since each vibrates at a particular frequency and amplitude. Any theory that locates sounds at their sources, and which ties the existence of a sound to the existence of vibratory activity in an object or objects, either by identity or supervenience, owes an account of transmission through barriers.

Even the wave view, on which sounds are waves, and other views that locate sounds in the medium face a dilemma. What, according to such theories, is the *source* of the sound? Is it the bell or the air–water interface? The preacher or the window? Each is in a sense the cause of the waves in the medium in which the sound is heard. An acceptable version of a medium-based view must acknowledge that sources are critical to auditory scene analysis and the individuation of sounds, and that we perceive locations in auditory perception, even if only the locations of sound sources. Just as the event theorist needs to say which events are the sounds, the wave theorist must say which things count as sources of sounds.

Though the problem is not unique to the event view, the event theorist owes an account of sounds and transmission.

The event theorist's options are: (a) deny there is a sound where transmission occurs and explain why the event view does not entail that there is; (b) accept that sounds accompany transmission events and reconcile this with the intuitive description of the experience.

Contrary to first appearances, option (b) is somewhat attractive. Suppose we accept that what occurs at a transmission interface or barrier is a sound—that the medium disturbance is a genuinely audible event. Granted, this event is caused in a very different way from the primary sound produced by an event such as striking a bell or shutting the door. It comes about because sound waves reach a medium with a considerably different density and elasticity and transfer a portion of their energy to that medium. The *sound transmission coefficient* of a medium is a constant that depends on the density, mass, and elasticity of the material. The portion of wave energy transmitted from one medium to another is a direct function of the degree of similarity between the sound transmission coefficients of the two materials. But what occurs at the transmitting surface or barrier is still notable from the point of view of the event theorist. The surface's activity is the proximal cause of motion in the surrounding medium that culminates in auditory experiences.

Two puzzling facts about auditory perception in such cases remain. First, transmitted sounds seem to come from where the original, or primary, sound event takes place. The sound of the preacher does not ordinarily seem to be at the window; it seems to be outside. If, however, a sound in fact exists at the window, it might preserve cues about the location of the initial sound. We perceive sounds as localized thanks to information contained in sound waves. When the waves from a source arrive at a surface or barrier, most of the information they carry about the direction and distance to that source remains intact when the abutting medium is set into motion. The result is a sound perceived to be located at the primary source, not at the transmitting barrier. If sounds occur where wave transmission does, these secondary sounds preserve their primary sounds' location information. This might, therefore, account for the misperceived locations of transmitted sounds.

Second, we classify sounds according to the types of events that we perceive to produce them. Yet sounds do not ordinarily appear to be produced by collisions between sound waves and surfaces or barriers. The primary sound, however, determines the purported secondary sound's qualities. Though less loud, the supposed secondary sound's qualitative and temporal profile would be roughly that of the primary sound. The secondary sound, which is caused by waves from the preacher's voice reaching a window, would thus resemble the primary sound. Given the preservation of location cues and the qualitative similarity of primary and secondary sounds, secondary sounds would be carriers of information about occluded primary sounds and would not be perceived to arise from events that occur at the window. This view attributes a great deal of illusion to auditory perception. The previously mentioned attraction is that it makes being a sound an intrinsic property of disturbance events. If being a sound is being a medium-disturbing event, then what makes something a sound is just that it is an occurrence with the physical makeup required of such a disturbing. There are not further extrinsic requirements stemming from the event's causal or temporal relations. If a window interacts with a medium to produce a disturbance, that window makes a sound no matter how its activities arise.

Ultimately, however, this response and the burgeoning world of sounds it proposes are unsatisfactory. It strains the imagination to suppose that a multiplying of sounds occurs each time sound waves travel across an interface or through a barrier. Our accounting should be more sober.

Suppose we deny that a sound occurs when a new medium is disturbed by a pre-existing sound wave. Option (a) suggests a different way to conceive of the perceptual situation. We say that the interface or barrier distorts our perception of the primary sound's location and qualities, not that we perceive a secondary sound, with its own location and set of qualities, which is caused by the primary sound. A single sound exists above the water or outside the window, but one may not have an ideal experience of that sound if impediments to perception intervene.

This picture is more accurate from a phenomenological standpoint. All indications are that we have a perceptual bias toward the locations of sound-generating events, that is, the everyday events that cause sounds to exist. Auditory perception makes us aware of sounds produced by sound-generating events such as doors shutting and ocean waves breaking. We hear the sound created by the striking of the bell above water, and the sound of the preacher proselytizing outside the window. Events of transmission occur when the waves from one sound event cause motion in an object or body that is passed on to another medium. We do not hear events of transmission or indeed anything at their locations when we hear a sound beyond an interface or through a barrier.

The language of this distinction suggests a theoretical solution consistent with the event view. To speak of a sound as *generated* by a source implies that the sound is caused by and distinct from the event that brought it about; but sound waves also are *generated* by sound-producing events. The idiom suggests that neither the sound nor the sound waves exist in any form prior to the event of generation. In contrast, the idiom of *transmission* suggests the passing along of a wave disturbance that already exists. The difference, however, is not merely relational. Indeed, the physics of sound wave generation differs from that of sound wave transmission: the physical characteristics of happenings that occur at the object or interface where wave generation takes place differ from those of sound wave transmission. The forms of energy transmission differ. During generation, something which is not itself a sound wave *produces* a sound wave; during transmission, sound waves *travel through* an interface or barrier as sound waves. Sound events involve the active production of pressure waves; transmission events do not.

When a transmission event involves a medium-disturbing or causes a wave disturbance of the sort that seemed to pose trouble for the event theorist, that transmission event depends for its existence upon a prior sound event. Nonetheless, the distinction between events in which sound waves are introduced into an environment and those in which sound waves are transmitted is natural. It is based upon the physics and on the events' roles in a regular causal network. The medium-disturbing events that are the sounds are the events in which a wave disturbance is *introduced* into the environment by the activities of some material object, body, or mass. Events in which a prior sound's waves are passed on or transmitted into a different medium are not in any ordinary sense sounds. An event of transmission must have a sound in its causal history, and a generated sound need not. Being a sound therefore is a matter partly of occupying a particular causal role.

But a central feature of the causal role distinguished by how we speak is supported by the physical distinction between generation and transmission. Sounds are events caused by collisions and strings, but sounds are not caused simply by waves passing through barriers and interfaces.

It is important to note that objects in the ordinary sense are not the only sound producers. For example, the column of air in an open tube or musical instrument often counts as the body or mass which itself vibrates and disturbs the surrounding medium in a wavelike manner. This case constitutes genuine sound wave generation on the part of a column or mass of air. Thunder is the sound of a rapidly heated body of air quickly expanding and contracting and thereby disturbing the surrounding atmosphere.

To summarize, sound-generating events such as cymbal collisions cause or generate sounds. Sounds are events in which a wave disturbance is introduced into a surrounding medium—that is, they are events in which an object actively disturbs a medium. The sound waves that result from the sound event, which also are in a sense generated, are transmitted through the surrounding medium and the materials and objects it contains. Events in which the motion of sound waves causes barriers and intervening matter to set adjacent media into motion are not events in which a disturbance is introduced into the environment. Such transmission events are among the inaudible activities of sound waves that carry information about the initial sound and propagate through various materials. Transmission events do not ordinarily give rise to sounds.

I say 'ordinarily' in light of the following sort of case. Suppose sound waves reach a barrier and induce vibrations in that object. The barrier might then itself generate a sound in addition to the sound whose waves induced the barrier's vibrations. This, however, is not an ordinary case of sound wave transmission. It is, instead, a case of *resonant sounding*. Resonating is sounding since the resonating object actively disturbs the medium, and does not merely passively transmit existing sound waves.

This account appeases tutored intuition. The problem of deciding which of multiple sounds we listen to when sound waves pass through an interface or barrier does not get off the ground. But the innocent picture according to which being a sound is entirely a matter of what happens near the surfaces of objects whose activities affect a medium is threatened. We must adopt a broader perspective that acknowledges the causal relations of several distinct kinds of events. This is no surprise given the organization of sound-related experience. Sounds furnish us with awareness of sound-generating events, which are of paramount interest for what they tell us about the world. They tell us such things as how the furniture is arranged and when it is being moved. Transmission events, however, enjoy little utility beyond what we learn through their effects on how we perceive the primary sounds they occlude; for example, when we perceive a sound as muffled, we learn that a barrier may intervene. Given, first, our interest in ordinary events that take place among material bodies, and, second, how these events are related to sounds, it is no wonder that the primary medium-disturbing events should be distinguished by audible qualities.

7.3 Destructive and Constructive Interference

As commonly demonstrated in high school physics classrooms, sound waves *interfere* with each other. Suppose you are in an anechoic room in which two tuning forks tuned to E above middle C are simultaneously struck. As you move around the room, there are places from which you hear the sound to be soft and places from which you hear the sound to be loud; there are places from which you hear neither sound.

This phenomenon occurs because at any time the total pressure at a point in the room equals the algebraic sum of the pressures of all the sound waves at that point. It is therefore possible, when sound waves are out of phase with each other, for the total pressure at some point or in some area to remain constant while separate sound waves pass through that point or area simultaneously. A listener positioned at such a point hears nothing. When sound waves cancel, the interference is *destructive*. Likewise, when the waves are completely in phase at a point, the total pressure varies with the sum of the components' amplitudes. The sound seems twice as loud as either tuning fork at these points thanks to *constructive* interference. Altering the phase or vibration characteristics of one of the tuning forks may result in *beating*, a periodic variation in perceived volume from a particular point.

Here is the problem. Take the example of complete destructive interference described above. The wave theorist can explain that you hear no sound from where you stand because the pressure is constant at that point and hence there is no sound. Of course, there are still in a sense two waves passing through that area, but their summed amplitude is zero. So, in a sense, there are two sounds at that point even though none is heard. The wave theorist does not escape entirely. If, however, by 'the wave' we mean something that depends only on the total pressure at various points, there is no wave and no sound at the point of interest.

By contrast, the event view implies that each tuning fork makes a sound even though you hear neither one from the point of interest. If sounds are not sound waves and the event view is correct, then you hear no sound at all when there are two. Is the gap a fault line in the event understanding of sounds?

Interference phenomena do not undermine the view that sounds are distal events. The interference arguments do show that waves carry information about sounds. Event theorists should not deny this when they say that the sound is not identical with the waves. Waves can be involved in the process by means of which a sound is heard without the sound's just being the waves. The event view provides an intuitive and compelling alternative to the standard account of destructive interference. The event view implies that there are two sounds—two events of a disturbance being introduced into a medium. The compression waves from these disturbances travel outward and may reach perceivers, where they cause perceptions of the original sound event. Waves obey the principles of interference, and if no variations in pressure exist, no sounds are heard. Ordinarily, a lack of pressure variations indicates the absence of sounds and sound sources. Complete destructive interference thus resembles the absence of sounds from an environment because factors conspire to create nodes where the pressure does not vary. These factors include the spatial arrangement of the two sources, the frequency and amplitude at which the sources oscillate, and the temporal relations among the activities of the sources, for instance, the phase difference of the sources. A perceiver located at a node will hear neither sound, and may believe that no sounds occur. This does not entail that the room contains no sounds. The observer simply is unable to perceive the sounds because of that observer's particular point of view. Likewise, listeners in a room with active noise cancellation, which works on principles of destructive interference, fail to hear any sounds.

That there are indeed two sounds can be confirmed in several ways. You can move to a point where one or the other sound is audible, move one or both of the sources so that the nodes are shifted, alter the phase difference in the vibrations of the two objects to remove nodes completely, or simply remove one of the sources to eliminate interference entirely. These exercises show that each tuning fork makes a sound that can be heard independently of the other in the right circumstances. Sometimes, however, another sound's presence can interfere with perceiving a given sound. Experience need not reveal from a particular vantage point all the surrounding environment's sounds. What we perceive from a very limited vantage point need not be the entire story about sounds.

The case of constructive interference is very similar. Because of the spatial and temporal relations among events of sounding, a perceiver in the right location may experience multiple sources to have greater loudness than any single source present. This again results from the additive properties of sound waves. It is less surprising that the subject's loudness experience should increase in the presence of two sources than that it should decrease, as in destructive interference. Beating is perhaps less intuitively comprehensible, but is also an explicable result of how the source events are arranged in time and space, and of the subject's vantage point on these events. From certain vantage points, it auditorily appears that a sound's loudness fluctuates.

Experiences related to interference phenomena can, on the positive side, provide information about the sounds actually present that would be difficult to obtain otherwise with unaided perception. Experience can tell us when two pitches are identical or if two sounds are exactly in phase when such facts are beyond the scope of ordinary perceptual discrimination. We tune a guitar by eliminating the beating between the open A string's note and the A of a pitch pipe. The two are in tune when the sound does not waver. Furthermore, one can tell that two sounds are exactly in phase when their combined loudness peaks.

7.4 The Doppler Effect

Explaining the Doppler effect is one of the event view's strong suits. According to wave conceptions of sound there actually are two Doppler effects. When a subject moves toward a stationary source, as in Figure 8, the subject experiences the sound as having a higher pitch than when that subject is stationary, a *subject-motion Doppler effect.* More wavefronts reach the organs of hearing per unit of time and thereby cause an experience as of a higher pitch.

The wave account explains the subject-motion Doppler effect nicely. The subject has an illusory or erroneous pitch experience due to motion relative to the stationary source. The event theorist can offer a similar explanation. The subject undergoes an illusory pitch experience because information about sounds is transmitted through the medium by waves. The sound of the stationary train's whistle remains the same despite the subject's motion.

Source-motion Doppler effects present the problem for the event theorist. When a source moves toward (or away from) a stationary

Figure 8. Subject-motion Doppler effect

subject, as in Figure 9, the apparent pitch shift differs from that which occurs when the subject moves at the same rate toward a stationary source.

Consider the following examples to illustrate the contrast.

Subject-Motion Doppler Effect

Suppose a subject moves at 100 meters per second toward a sound source that emits 1000 waves per second (1000 hertz). Since the speed of sound waves (v) in air is 340 meters per second, the wavelength (λ) in air of 1000-hertz waves is 0.34 meters ($\lambda = v/f$). So in 1 second the subject will pass 1000 waves plus the additional

Figure 9. Source-motion Doppler effect

waves that occupy the distance she travels, 100/0.34 waves. The resulting apparent frequency is 1294 hertz.

Source-Motion Doppler Effect

Suppose a source emitting 1000 waves per second moves toward a stationary subject at 100 meters per second. Since the speed of sound in air is 340 meters per second, after one second 1000 waves will occupy a distance of just 240 meters. This amounts to a wavelength of 0.24 meters and, thus, a frequency of 1417 hertz.

According to a wave-based theory of sounds, the sound itself changes in character with source motion since the waves in the medium have a different frequency. Sound waves actually are more tightly bunched as a result of the source's motion through the medium. The subject thus enjoys a veridical experience of a sound with qualities different from those of a stationary source.

The event view, which identifies the sound with the disturbing event, cannot readily explain the apparent change to sound quality in terms of wave bunching. The vibratory motion of the surface that defines the event's location determines the pitch of the sound—but that vibratory activity remains constant despite source motion. Event theorists have two options. We can try to accommodate a difference in pitch in the source-motion Doppler, or we can deny that either type of Doppler effect amounts to a genuine difference in the sound's pitch. The latter subsumes both types of Doppler to perceptual illusion. I opt for the latter.

Why? Suppose that the pitch of a sound does change with source motion. We might say that the disturbance event is different in this case because the interaction between source and medium differs as a result of source motion. It is reasonable that the event itself is different because its effects are different.

We should resist this for at least two reasons. First, though different causes and effects may be sufficient for there to be a different particular event, in this case they are not sufficient to characterize the different qualitative character of the disturbing event. That is, pitch is a *quality* of sounds, and different effects do not give reason to attribute a difference in pitch to the disturbance event without sufficient differences pertaining to the sound event itself. Since the disturbance event, and in particular its vibratory motion, does not change in relevant respects when it is in linear motion, we should not say that its pitch changes.

Second, even if we did say that the sound event changes with source motion, it would follow that in source-motion Doppler cases the sound has a different qualitative character at different places where the object interacts with the medium. That is, the pitch of the sound at the front of the whistle would differ from the pitch of the sound at the rear of the whistle. If motion suffices for pitch change in one direction, it suffices for pitch change in the other. So the character of the medium-disturbing varies with location along the surface of interaction. It is hard to see how linear motion alone could justify attributing different pitch qualities to different spatial parts of a sound event.

We should for these reasons say that the sound event maintains a stable qualitative character through changes in its position relative to observers—whether due to its own motion or to the motion of observers. The source-motion Doppler is thus an apparent shift in the pitch of a sound due to wave phenomena.

Is it independently plausible that both sorts of Doppler effect are cases of perceptual illusion? Suppose that, despite how things might appear auditorily, a tendency toward a form of constancy judgment for pitch exists in Doppler cases. Such constancy judgments should be present in both the subject-motion and the source-motion Doppler. That is, suppose that we withhold endorsement of the claim that a sound shifts in pitch when either a subject or a sound source is in motion with respect to the other, and instead judge that the sound maintains a constant qualitative character despite the motion of the subject or of the source. It is plausible that such constancy judgments actually occur and can be elicited when we hear sounds with moving sources. I do not actually think that a train's sound rapidly changes from having an elevated pitch to having a diminished pitch when it speeds past the platform. Rather, I am inclined to think that I hear the train's sound without distortion when we both are stationary relative to each other. There are not two sounds of the moving train to be heard by passengers and bystanders, respectively. If this is genuinely plausible, then we have good reason to count both sorts of Doppler effect as perceptual illusions, just as we do when visually perceived qualities are Doppler 'shifted'. Despite how things might visually appear, once we take into account the motions of celestial bodies, we do not judge that the colors of the stars are the ones they visually seem to have. Thus, the event view implies what is already reasonable, that relative motion does not suffice to change the audible qualities of a sound.

The event view therefore yields a unified explanation of source-motion and subject-motion Doppler effects. Both source motion and subject motion produce illusions of altered pitch thanks to how waves transmit information about sounds and excite auditory experiences. In neither case does a sound alter its qualities due to relative motion of source and subject. Rather, a sound merely seems to have altered its pitch thanks to such relative motion. The event view thus captures in a unified explanation the way experienced pitch depends upon the motions of subjects and sound sources. In fact, it furnishes a powerful framework for understanding the conditions under which auditory illusion occurs because it delivers standards concerning when hearing is veridical, standards that are independent from fluctuations to hearing that occur because of changes in the local acoustic environment. Despite auditory appearances, distortions to waves stemming from transmission, interference, and Doppler effects lead to perceptual distortions, and do not constitute differences in the world of sounds.

In the upcoming chapters, I continue the discussion of sound-related phenomena by exploring and developing an account of echo experiences produced by encounters with reflected sound waves. The central question for the theory of sounds and sound perception is whether echoes are genuine sounds or, instead, a variety of perceptual phenomenon whose appearance results from the behavior of sound waves. The case of echoes and the Doppler effect have more in common from a perceptual standpoint than one might expect. The perceptual experiences associated with Doppler shifts and apparent echoes, as well as with interference and transmission phenomena, are predictable perceptual effects explained by the wave-based transmission of information about sounds. Perceptual strategies and rules for identifying, individuating, and tracking sounds are not infallible. At times wave phenomena lead to experiences that are to some extent illusory. But such illusions are instructive. By recognizing the illusions we in fact fall prey to and the circumstances that produce them, we learn not just about the proximal causes of auditory experiences, but also about the events and objects that constitute the primary targets of our perceptual interest in an environment.

8

The Argument from Echoes

Suppose you are at a fireworks display. You stand in an open field with a single brick building behind you. A colorful bomb's recognizable boom follows its visual burst, but a moment later the boom's echo sounds at the brick wall behind the field. You have just heard a *primary sound* followed by its *echo*.

I have argued that sounds are events. In particular, a sound is the event of an object or interacting bodies disturbing a surrounding medium in a wavelike manner. According to this event conception of sounds, the sound of the firework just is the disturbance event that occurs where the explosion affects the surrounding air. A theory of sounds should provide an account of echoes and echo experience. The appearance of echoes threatens to undermine the event conception of sounds because the echo seems to exist well after, and at a distance from, the medium-disturbing event.

If the primary sound and the echo one hears are the very same sound, that sound appears to exists after the medium-disturbing event is complete. The worry is that the case of echoes demonstrates that sounds persist through the time between hearing a primary sound and hearing an echo. If the existence of echoes entails that sounds themselves exist long after their sources have ceased to exist, then sounds are not the events that I have suggested. Furthermore, if the case of echoes demonstrates that sounds as a rule outlast their sources, then since primary sounds and echoes seem to be heard at different locations, the case of echoes provides fresh fuel for the view that sounds travel. If the appearance of echoes entails that sounds persist beyond medium-disturbing events, may be reencountered at later times during this continuous career, and may travel independent of their sources, the case of echoes establishes that sounds are not the distal events with which I have identified them.

An alluring form of argument involving the case of echoes, based on these worries, has been leveled against the view that sounds are events. In this chapter, I first argue that although the claim that sounds travel appears attractive in the case of echoes, it in fact gains no further support from the case of echoes. There simply is no good non-question-begging argument for the claim that sounds travel. However, I also explore whether echoes demonstrate, independent from the claim that sounds travel, that sounds persist between primary sound and echo experiences. Once we articulate the sense in which experiencing an echo is reencountering a particular sound, we shall see that the case of echoes in fact presents grave trouble for those who hold that sounds persist between primary sound and echo experiences. That claim is incompatible with views about the temporal features of sounds that we have reason to accept on independent grounds. Since it is not obligatory to accept that sounds persist between primary sound and echo experiences in order to provide a compelling account either of the content of echo experiences or of the ontology of echoes, this sets the stage for a novel understanding of echoes and echo phenomenology according to the event conception of sounds. I present this in the next chapter.

8.1 Do Echoes Show That Sounds Are Not Events?

Matthew Nudds has recently offered the following argument that sounds are not events.

One of Newton's minor triumphs in the *Principia* was a derivation of the velocity of sound. To test his derivation he measured the time for an echo to return from the end of a colonnade in Neville's Court of Trinity College. Lacking anything resembling a stop watch, he adjusted a pendulum to swing in rhythm with the echoes of successively made sounds (Westfall 1980: 455–6). Newton used the pendulum to measure the time it took for a sound to travel from its source down the colonnade to a wall and then back again; he heard the particular sound that he had produced a moment earlier reflected back to him. Here we have a simple example of a sound existing even after the event which produced it ceases (we can suppose) to exist, and which moves independently of whatever produced it. It is also an example of a single sound being heard more than once: Newton re-encounters a particular sound, as he must do if he is to time its journey up and down the colonnade. Sounds, then, are particulars. Normally we think of events as things that, unlike objects, cannot be re-encountered. The fact that we can re-encounter sounds suggests that they are perhaps best thought of as kinds of objects, rather than as events; either way, not as properties. (Nudds 2001: 221–2)

What I shall call the *argument from echoes* is this:

(1) Newton measured the speed of a sound by measuring the time it took for the sound to travel down a colonnade and back.
(2) Hearing an echo is reencountering a particular sound.
(3) Events, unlike objects, cannot be reencountered.
(4) Therefore, sounds are object-like particulars and not events.

If this reasoning is cogent, an echo is a particular sound at a later stage of its continuous career, after it has been reflected.

8.2 Sounds Do Not Travel

Consider the following two alternatives to (1).

(1′) Newton measured the speed of *sound waves* by measuring the time it took for him to hear an echo after hearing the primary sound.
(1″) Newton measured the speed of *sound waves, but not of a sound* by measuring the time it took for him to hear an echo after hearing the primary sound.

Alternative (1″), which I accept, is a strong replacement for (1) that does not entail (2). Alternative (1′), however, is uncontroversial. But (2) does not follow even from the weaker (1′). Hearing an echo may not be reencountering the same sound at a later stage of its career, even if we owe the episodes of hearing to the same sound waves. The conclusion that sounds are not events does not follow from the argument reconstructed with (1′). The argument from echoes against the event view requires nothing short of (1).

Though I have argued that sounds do not travel, it is certainly widely accepted that sounds do travel. But why? In contrast to (1′), (1) is not obvious. By now the following theme is well rehearsed: sounds usually are heard to travel only when their sources do. If sounds seemed to you to emerge from their sources and to speed through the air toward your head, we would send you to a specialist.

Maybe echoes are special. Perhaps the argument from echoes reflects the fact that hearing a sound followed by its echo forms the basis of a strong intuitive case for the belief that sounds travel. After all, the argument is not meant to prejudge the theoretical question of what a sound is. Rather, it aims to enlist conceptions about sounds and echoes implicit in ordinary beliefs to show that sounds are not events. Indeed, the perceptual facts in the case of echoes ground a tempting inference to the effect that sounds travel. But the transition from experience to belief is not a matter simply of lending credence to that which is already apparent in experience. Believing that sounds travel requires more than just accepting the way things appear to you. Take a sound, *s*. *That* s *travels independently of its source* is not ordinarily part of the content of an auditory experience, and it does not arise as part of the content of the experience when a sound is followed by an echo. One might think that this observation is all that is required to show that sounds do not travel, if one thinks that a sound cannot have properties beyond those it is experienced to have. This is too easy. I leave it open that sounds can have properties we are unable to detect with our ears alone, and that there can be sounds humans are unable to hear. Sounds may indeed move through space; whether they do is a matter of what the correct theory of sound is. This is not to say that our goal should not be a theory that coheres with the contents of experience—sounds are, after all, among the things we hear. We would like it if perception served as a reliable guide to reality, but whether it does with respect to sounds is a matter to be decided by a variety of concerns. So perhaps the case of echoes presents further reasons that make the belief that sounds travel intuitively appealing.

We do assume that each sound is generated by some event that is not a sound. It appears to be a working assumption of the auditory perceptual system that each sound must be produced by some event in the environment (see Blauert 1997, especially Chapters 2 and 3). Furthermore, it is an introspectible part of experience when hearing a sound that the sound is a sound *of* something going on in the surroundings. The belief is plausible and sustainable since empirical evidence repeatedly confirms that sounds are caused by happenings that take place among material things—we are hard-pressed to find sounds for which investigation reveals no source events.

The fact that echoes strike us as causally related to events that occur elsewhere—in fact, as *generated* by distal events—establishes a connection to what we take to be the source of the primary sound. The echo of the firework's blast strikes us as being connected to the explosion that generates the primary sound. Though the primary sound and the echo seem distinct, they appear to share a common causal history.

But since no source appears to exist where the echo is heard to be, from (i) the qualitative similarity of primary sound and perceived echo, (ii) the time delay between the experience of primary sound and of echo, and (iii) the shared causal history of the primary sound and echo, it is natural to conclude that what we think of as the primary sound and the echo are the very same sound. Once we overcome the apparent distinctness and identify the primary sound and the echo *in thought*, two facts—that the primary sound is heard to be near the source at one time, and that the echo is heard to be at some distance from the source slightly later—may encourage us to believe that the sound has traveled.

Note that forming this belief requires us to ignore or implicitly reject the possibility that two different sounds (with different locations) are generated by the same event. This rejection itself requires accepting either that a sound cannot be generated at a distance from its source or that each sound has a unique generating event. However difficult it may be to motivate the latter, the former is on its face quite plausible.

Even after accepting all of that, you are entitled only to believe that the sound traveled from the perceived location of the primary sound to the perceived location of the echo. Nothing about the naïve perceptually based way of thinking about the situation allows you to go beyond this to the belief that sounds travel as waves do, that is, that the sound traveled from its source to the reflecting surface and back to you. The case of echoes adds little to the case for (1).

The force of (1) must come from one of two sources: an appeal to science or an inference that fills the gap in the above line of thought.

Consider the first. It is common knowledge that information about sounds is transmitted by pressure waves in a medium. To accept that sounds travel because pressure waves travel, however, is to complacently accept a caricature of scientific doctrine. Implicit disagreement about the nature of sounds exists even within scientific communities. Frequently researchers analyze and divide the terrain, and avoid commitment on the more general question. When they do commit, physicists most frequently speak of sounds as waves or as purely mental phenomena, but perceptual researchers tend to speak of distally located *acoustic events* in addition to sound waves and auditory experiences (see, for instance, Bregman 1990, Luce 1993, Carlile 1996, Blauert 1997, and Gelfand

1998 for illustrations of this contrast). Science uncovers the relevant causal processes between environmental happenings and auditory experiences, but it need not legislate on the ontology of sounds. Part of the business of science is to explain by means of discovering and articulating causal connections. Though sounds cause auditory experiences, and pressure waves cause auditory experiences, the identification is nevertheless unwarranted. To conclude, from '*a* causes *c*' and '*b* causes *c*', that '*a* is identical with *b*' is fallacious; from 'guns kill people' and 'people kill people', we cannot conclude 'guns are people'. Given that sounds and sound waves differ in significant causal and perceptible respects, it begs the question to conclude on the basis of science that sounds travel.

How about the second? Consider the following inference that attempts to bridge the gap between sound sources and perceivers.

(0) The sound was generated over there; I am here; since I heard the sound it must have traveled to reach me.

But since the sound itself is perceived to be *stationary* and *distant*, this inference is either question begging or fallacious. It begs the question to assume that the sound is identical with the waves, and therefore travels. It is fallacious to conclude that since distance separates perceivers from sound-generating events, sounds must travel. Consider the following correlate of (0).

(0′) I heard the sound; then I traveled faster than the speed of sound waves, stopped, and heard it again; so, the sound must have traveled.

To explain why one hears the sound a second time does not require that the sound travels. Sounds need not travel since that by means of which perceivers hear them—sound waves—can move independently of sounds.

Premise (1) of the argument from echoes is seriously undermotivated.

8.3 Reencountering Sounds

Consider, however, the argument's remaining premises, (2) and (3). Although the truth of (1) supports (2), the argument from echoes as it stands does not require premise (1) and the claim that sounds travel for its validity. To establish the conclusion that sounds are not events it suffices to establish that hearing an echo is reencountering a particular sound and that events cannot be reencountered in this way. A reformulated argument from echoes threatens to undermine the identification of sounds with events.

Do echo experiences result from encounters with already-heard sounds? That hearing an echo after hearing its primary sound is reencountering the same persisting sound particular finds little support in perceptual experience. As I pointed out above, primary sounds and their echoes seem to be distinct. A single object perceived at different times does not. Hearing an echo is unlike reencountering a person you have met before, and it is unlike glimpsing someone carrying home the vase you saw earlier in a store window, even though the echo has an equally rich qualitative signature. The perceptual experience of an echo does not bear the marks of *recognition* and *reidentification* that characterize the experience of material objects and continuants with relatively stable qualities. We find it reasonable to count echoes and primary sounds as distinct sounds before doing philosophy. But to count earlier and later states of persisting things as distinct objects is counterintuitive. Not only does the continuous experience of an ordinary sound differ from that of an ordinary object in that the object but not the sound seems wholly present at a given time, but the discontinuous experience of a primary sound followed by an echo differs from that of reencountering a persisting object. The echo experience, but not the later encounter with an object, seems to have as its object a particular distinct from one perceived earlier. Again, in the debate about whether each material object is a series of suitably related but distinct *temporal parts* or a single thing that undergoes changes in properties through time, that is, whether things *perdure* or *endure*, it is the latter endurantist view that appears natural to common sense and comports with perceptual experience. The perdurantist about objects must motivate the view with philosophical considerations. The view that echoes are distinct sounds appears not to require similar philosophical motivation. Premise (2) therefore finds little support from auditory perceptual experience.

Matters get worse for the reconstructed argument from echoes. Given a plausible thesis concerning sounds, accepting an interpretation of 'reencounter' in (2) and (3) that makes the argument valid leads to trouble. As I argued in Chapters 2 and 3, sounds extend through time. Much as concrete objects occupy space, sounds occupy time. Objects fill space, and sounds take time. Each sound has a duration: it begins at one time; it ends at another (if it ends at all).

A number of considerations already canvassed support the following *temporal thesis*.

(5) Sounds have duration.

First, sounds are not, or certainly do not seem, *wholly* present at each moment at which they exist. Compare sounds to objects. It is plausible that all that is required to be a particular object can be present at a moment and, thus, that a particular object can be wholly present at each moment at which it exists. At any particular moment, however, only a portion—a temporal part—of a sound is present. That is, speaking strictly, rather than a sound in its entirety, only a part corresponding to a property profile at a time is present at each moment during which the sound exists. Second, sounds survive change. A given sound may continue though its properties shift markedly. A sound may involve a great deal of qualitative change. Third, having a qualitative profile *over time* is central to the identity of a particular sound. Consider a spoken word or a particular squawk from a raven. Each is constituted from qualitatively different stages; the particular pattern of changes each exhibits through time determines the kind of sound it is;

and sounds whose profiles differ over time belong to different sound types, whether or not they incorporate similar stages. These considerations and the temporal thesis (5) they support are independently plausible statements that express judgments grounded by our intuitive conception of sounds.

The argument from echoes relies on the claim (3) that events are not the sorts of things that we can reencounter. We need a working conception of what it is to encounter something in the relevant sense. I shall continue to assume that to encounter something is just to enjoy perceptual awareness of it. To the extent that we can speak meaningfully about encountering an event, nothing more is required than to be in a suitable relation of perceptual awareness to it. An alternate sense of 'encounter' means roughly the same as 'bump into'. This sense seems appropriate if sounds are the traveling causes of auditory experiences, as premise (1) would have it, but applies paradigmatically to objects and not to events. Since the argument from echoes employs the fact that sounds can be reencountered to show that they are not events, this must not be the sense of 'encounter' in question. Otherwise, to show that we bump into sounds would suffice to show that they are not events. But the argument seems to be: we certainly can encounter events; we just cannot *re*encounter them. If the argument turns on something peculiar about echoes, 'encounter' must mean something like 'perceive' or 'enjoy perceptual acquaintance with'. Though to make the notion line up with ordinary usage might require adding some proximity requirement, perceptual acquaintance suffices for what follows. The argument must be that while we can reencounter sounds in the form of echoes, we do not normally think that we can reencounter, that is, reperceive, events.

Sharpening the sense of 'reencounter' in this way is not enough to make the argument from echoes valid. Suppose I wake up to hear the loud, high-pitched beginning of the local emergency siren's wail. I then descend into the silent basement for two minutes, after which I emerge to hear the nearly completed sound's fading low-pitched moan. I experience a different part of the sound upon each hearing. Thus, we can reencounter a particular sound by encountering different parts of the same temporally extended sound. Consider the following replacement for (2).

(2′) Hearing the end of a siren one heard earlier is reencountering a particular sound.

Events, however, often are reencountered in this way. I might watch the first part of a high dive, turn away, and then look back to see the splash. Premise (2′) is consistent with sounds being events. The argument reconstructed with (2′) does not establish that sounds are not events because on this understanding of 'reencounter' (3) is false.

What the argument from echoes needs in order to show that sounds are not events is that we can reencounter the *very same parts* of a particular sound on separate occasions. From this it follows that if we can encounter all the parts of a sound on each of two different occasions, we are capable of reencountering the sound *in its entirety*. The import of (2) must be the following strengthened claim.

(2″) Hearing an echo is reencountering a particular sound in its entirety.

The claim of (2″) must be the one intended to ground the argument that sounds are not events. We do not ordinarily think of events such as particular closings of doors and birthday parties as things that can be reencountered in their entirety since they occur at specific times and places. Halloween is a different event each year that belongs to a class of like events each of which occurs on some 31 October. While we take for granted reencounters with persisting objects, to reencounter a particular Halloween would take a special trick, such as time travel. Of course you might reencounter a sound if you traveled back in time, but that is not ordinary.

Another such trick in the case of sounds would be to travel faster than the speed of sound waves. Suppose you perceive a sound the first time, then speed past the waves, stop, and enjoy a second experience of the sound. You reperceive the sound, even if it is an event, because you overtake and happen again upon its traces. This is not ordinary, and does not represent a usual way of reencountering events. For the argument from echoes to be valid, the import of (2) and (3) therefore must be the following.

(2‴) Newton, upon hearing the echo, reencountered in its entirety and without special tricks the very same particular sound.

(3‴) Events, unlike objects, cannot be reencountered in their entirety without special tricks.

The trouble comes when we accept (5), the temporal thesis, while accepting a version of (2) that implies (2‴). From (2‴), in the case of echoes a particular sound can without special tricks be encountered in its entirety on two separate occasions. From the temporal thesis (5), each sound has duration—it begins and ends. To recognize the problem, consider hearing a particular sound and its echo. It is an experiential fact that between t_1 and t_2, call this interval $\mathbf{t}_p$, we experience the primary sound. Likewise, the echo is heard during some later interval $\mathbf{t}_e$, which begins at t_3 and ends at t_4. By (2‴), the sound heard during $\mathbf{t}_e$ is the same particular sound heard during $\mathbf{t}_p$. Now, since no special tricks are required to reencounter the sound, and since temporal continuity is necessary for the identity of a particular, the sound must persist or exist continuously between $\mathbf{t}_p$ and $\mathbf{t}_e$. It follows from this that the reencountered sound is perceived to begin and to end multiple times during which it continuously exists. Something cannot, however, continue to exist after it has ceased to be. Something cannot begin or end multiple times. Therefore, either the temporal thesis (5) is false and sounds do not have the durations they appear to have, or hearing an echo after hearing a primary sound is not reencountering a particular sound that persists while unheard. Without an argument for the claim that sounds travel or for the claim that sounds are reencountered in the required sense, we should preserve the temporal thesis and reject the claim that sounds must persist between primary sound and echo experiences. The reconstructed argument from echoes fails because (2‴) is incompatible with well-motivated claims about essential temporal characteristics of sounds. Furthermore, as I shall argue below, (2‴) is dispensable.

A proponent might nonetheless attempt to come to the defense of the argument from echoes as follows: When you perceive a sound to begin and end, you perceive the sound's spatial boundaries, its 'front' and 'rear', as it travels past. We mistake the beginning of our experience of the sound for the temporal beginning of the sound, and the end of our experience for the temporal end of the sound. The sound, properly speaking, begins and ends only in the spatial sense. It has a front and a rear, the experiences of which lead us to believe that the sound begins to exist and then ceases. By experiencing the spatial boundaries of a sound multiple times, the reencountered sound can seem to begin and end entirely on separate occasions.

The explanation of our perceptually based confidence in the temporal thesis adverts to encounters with the spatial boundaries of sounds as they travel past. These encounters correspond to perceived beginnings and endings of sounds, while only our experience of the sound properly can be said to begin and end.

As I argued previously in Chapter 3, from this conception several unsavory consequences follow that together suggest we are on the wrong track if our goal is to provide a phenomenologically apt theory of the objects of auditory experience. It seems patently false—a violation of the transparency of experience—that we mistake the duration of an experience alone for the duration of a sound or for the duration of a sound-generating event in our environment. I have worked to undermine the claim that sounds travel independently from their sources in part because the conception of sounds that it underwrites is so unattractive. To believe that hearing an echo is reencountering a persisting sound only amplifies the counterintuitive consequences.

Either this conception paradoxically implies that events that cause sounds—sound-generating events such as firework explosions and door closings—are not events, or, once again, it implies that we enjoy veridical perceptual awareness of the durations of such sound-generating events by means of illusory awareness of the durations of sounds. Consider the first disjunct. Since we reencounter the sound when hearing an echo, and since perceiving a sound is a way of perceiving a sound-generating event, we reencounter the sound-generating event by hearing the echo. Since events cannot be reencountered, events that cause sounds are not events. That is absurd. So consider the second disjunct. Perhaps hearing an echo is a special trick for reencountering an event. In that case, however, hearing an echo still is a means to hearing the duration of a sound-generating event. Though the experience of an echo's duration is illusory, one's experience of the sound-generating event's duration is veridical. The experience of a primary sound followed by an echo doubly affords veridical awareness of a sound-generating event's duration by means of an illusion.

Moreover, it follows that both the structure of our perceptual judgments and the resulting conception of how sounds exist unperceived are faulty. We think that sounds can exist unperceived in the following sense. When a car passes me and goes out of range, I cannot hear its sound. The sound still exists, but I cannot perceive it given my location and auditory capabilities. We do not ordinarily think that a sound exists unperceived in the sense that it exists after we have heard it and after the sounding has stopped. But on the proposal being entertained, sounds exist long after the events that produce them do, and, indeed, they continue to exist well after we hear them. So, even though you think the sound of a spoken word no longer exists after you hear its last syllable uttered, it goes on existing for as long as the corresponding waves travel through the medium. This consequence is strikingly at odds with common and well-founded implicit judgments about the persistence conditions of sounds. We are better off accepting the temporal thesis that sounds often have the durations they seem to possess than multiplying misjudgments. This requires rejecting the argument from echoes.

8.4 Trick Reencounters

It is more difficult to motivate the claim that sounds travel than you would expect from the way we sometimes talk. Echoes offer little to further the case. The claim that sounds persist between primary sound and echo experiences underwrites a conception of sounds that conflicts with central elements of the way we perceptually individuate and reflectively judge sounds to be. The argument from echoes therefore fails to demonstrate that sounds travel or that sounds are persisting particulars that may be reencountered without tricks and in their entirety. It therefore fails to show that sounds are not events.

However, even if the case of echoes is compatible with the claim that sounds are distal events, echoes still appear to be distinct sounds located at reflecting surfaces. But since the brick wall, for instance, merely reflects sound waves and does not actively disturb the surrounding medium, the version of the event view I have proposed appears to have no sound to identify as the echo. If sounds do not travel, where is the echo? If sounds do not persist between primary sound and echo experiences, what explains the later appearance of echoes? In the next chapter, I propose and develop an account of echoes and echo perception that addresses these remaining worries while preserving audition's fidelity to the temporal thesis.

I shall argue, perhaps surprisingly, that to enjoy an echo experience is to reexperience a primary sound, and that the apparent echo just is the primary sound. This is perhaps surprising since I do not accept that sounds travel, and I do not accept that sounds must persist between primary sound and echo experiences. One need not, however, accept that sounds are particulars that travel or persist to identify the apparent echo with the primary sound. According to my account of echoes and echo perception, hearing an echo is hearing the primary sound, but with distortion of place, time, and qualities. Hearing an echo thus is like seeing an object with a mirror. Sounds need not travel since that by means of which we hear them does. The echo experience and the primary sound experience seem, nonetheless, to have distinct objects because we ordinarily perceive entire events only once. Hearing an echo, therefore, is reencountering a sound with the help of a special trick; the reflection of sound waves from a surface serves as the occasion for the trick. You reperceive the primary sound event because its traces rebound and return.

Considering the issues brought to light by the case of echoes improves our understanding of the nature of sounds and serves as part of the positive case for the event view. The event theory grounds a complete account of the metaphysics of sounds and the phenomenology of auditory experience.

9

Echoes

9.1 The Problem of Echoes

An account of echoes and echo perception should say what an echo is, explain the distinctive phenomenology of echo experiences, and respect scientific descriptions of the physical and perceptual processes involved in hearing echoes.

Suppose again that you are standing in a large open field with a brick wall behind you, and that you hear a firework followed by its echo at the brick wall. When you hear the sound of the firework and its echo, certain features of the experience are distinctive. First, you hear the echo after you hear the primary sound. Next, you hear the primary sound to be located near the explosion itself, but you hear the echo to be located near the reflective brick wall. Though the echo appears to be a distinct sound, investigation (perhaps mere visual awareness) reveals no sound source at the brick wall. Nothing at the reflecting surface generates the apparent sound. Finally, the echo and primary sound ordinarily appear to have similar qualities and duration. The degree of distortion depends on the qualities both of the sound and of the reflecting surface.

When the firework explodes, it disturbs the surrounding air, and pressure waves travel outward toward you and toward the brick wall. When the waves reach you, they contribute to your experience of the primary sound. As waves reach the brick wall, an elastic collision takes place and the wall redirects the waves. These redirected waves reach you and produce the experience as of a second sound distinct from but somehow related to the initial sound. This time gap is essential to enjoying the distinctive multisound echo experience. When secondary sound waves arrive at the ears less than about fifty milliseconds after primary sound waves, the result is an experience as of a single sound located between the two wave sources. Between roughly fifty milliseconds and two seconds, the result is an experience as of a primary sound and a distinct but somehow causally related secondary sound or echo. When the arrival delay is greater than about two seconds, the experience is as of two separate and entirely unrelated sounds. I will focus on the general case of reflected sounds and hope to explain the distinctive multisound echo experience in purely perceptual terms. The account of echo perception that I offer is consistent with the psychophysical facts and makes the phenomenological differences that attend different time delays depend primarily on features of the auditory perceptual system.

The trouble for the disturbance event account is that a mere elastic collision occurs at the brick wall. The brick wall does not vibrate or actively introduce a disturbance into the surrounding medium. It simply gets in the way of sound waves that already are present. The case of sound wave transmission through barriers, discussed in Chapter 7, taught that passing along or redirecting a pre-existing sound wave is not sufficient for producing a sound. The event view appears to have no disturbance event to identify as the echo.

If sounds are particulars that persist between primary sound and echo experiences, and travel through the medium, a simple resolution exists. Hearing an echo after a primary sound is hearing the very same sound particular at two different stages of its continuous career. This, however, is incompatible with the event view, according to which sounds are events that do not outlast and whose locations are stationary relative to their sources.

9.2 The Solution

The echo just is the primary sound. According to the event view, the apparent echo is the original or primary sound event perceived with distortion of place, time, and qualities. The illusion is caused by the behavior of sound waves and the way in which sounds are perceptually localized. Sound waves, which transmit information about sounds, are the proximal causes of auditory experiences. Their direction of onset determines the perceived location of a sound, and their rate of travel results in the delayed echo experience. Filtering and dispersal at the reflecting surface account for differences in the perceived echo's qualitative profile.

Hearing an echo, then, is hearing a primary sound. You hear the primary sound, then after a short delay you hear the sound *qua* echo. What you hear in each instance is the very same sound—the primary sound. The primary sound is not an object-like particular that travels through space. It need not persist between the subject's primary sound and echo experiences. The sound is an event that occurs only once at the location of the sound-generating event, which does not ordinarily outlast that event. The traces of the primary sound—its sound waves—travel and encounter reflecting surfaces. When the waves return, an appropriately situated subject has an experience as of an echo, a seemingly distinct though causally related sound located in the direction of the reflecting surface. The subject, however, hears only the original sound on each occasion. Strictly speaking, the distinction between primary sound and echo is entirely perceptual and depends upon wave arrival time delays at a subject. The interval between wave arrival times from the very same sound determines whether reexperiencing it results in a distinctive multisound echo experience. The apparent causal relation makes the experience distinctive. The causal relation, however, need not appear to hold between the two sounds. The primary sound and echo instead may seem to share a common causal source.

This account is analogous to a plausible treatment of seeing objects with mirrors. Mirrors facilitate our seeing the very objects and events that occur in front of them, albeit with distortion of place and perhaps qualities, as with an aged or funhouse mirror. Likewise, reflecting surfaces allow us to hear the very sounds that occur in front of them, albeit with distortion of place and time. But just as there is no new object that you see when you look in a mirror, there is no new sound that you hear at a reflective surface.

Hearing as of an echo after first hearing the primary sound is, on this account, an unobjectionable sort of reencounter with the very same sound. The sound event occurs once, say between t_a and t_b. You experience it once between t_c and t_d and again between t_e and t_f because the waves it creates return. The sound neither travels nor returns to you; you experience the same distal event twice because of the way the event's traces travel. The situation is something like a case discussed in the last chapter. Suppose you hear the sound of the firework. You then travel faster than the speed of the sound waves, overtake them, and halt. You now hear the sound again—it seems to be in the same place it was before. We need not say that the sound travels, only that the sound waves travel. Because information about sounds is transmitted through a medium by means of relatively slow waves, you are lucky enough to enjoy the same sound event twice. The medium-disturbing event you hear when you hear the sound for the second time is the very same disturbance event you heard earlier. Echo perception is similar. A reflecting surface, however, saves you the trouble of supersonic travel. You pay the price with distortion of location.

Hearing an event that is past is thus like seeing an event that is past. When you see a supernova from across the galaxy, you see that event as it happened long ago. But you experience it to be present—to be taking place *now*. So your experience includes a temporal illusion. Now, suppose there were big mirrors in outer space. You could then see the very same earthly event twice: once when it happens and once after its traces are reflected. I could watch the Olympic Marathon finish on 24 August 2008 and then watch it again on 24 August 2009 in a mirror located one-half light year away. If the mirror was big enough, I might even think there was a race being run on a far-off planet that looked remarkably like Earth. The case with echoes is a less exaggerated parallel. Hearing an echo does not involve such great distances.

Still, the echo experience and the primary sound experience seem not to have the same object. If the apparent echo is the same sound as the primary sound, why do we not recognize it as such? Of course, we sometimes fail to recognize an object as one we have seen before, but when the object is qualitatively similar on both occasions this requires special circumstances. Barring failures of recognitional capacities, including memory, we can usually be made to recognize persisting objects.

The apparent distinctness of echoes from primary sounds is due to the nature of events and to how we conceive of them in contrast to objects. If a primary sound and an echo were one object experienced at different times during a single continuous career, we would expect ordinary object recognition and reidentification to occur, given the qualitative similarity of the echo to the primary sound. With objects we count on this sort of recognition to ground the perceived continuity of the material world. *Capgras syndrome* is one form of *delusional misidentification syndrome* in which patients suddenly begin to believe that people and objects familiar to them have been replaced by exact qualitative duplicates. This failure of perceived continuity is notable and debilitating (see Breen et al. 2000). Events and time-taking particulars, however, are tied to a specific time and place at which they occur. Though the 2006 World Cup final might have been located at any of various times and places, it in fact occurred 9 July 2006 at Berlin. That very event cannot occur again or elsewhere. And we implicitly recognize this: similar events experienced at different times and places are taken to be distinct events. So, if we happen to perceive the very same event over again, it should seem like a distinct event. What about watching an instant replay? The same event perceived a second time does not seem distinct. But this simply is a matter of habituation—we are used to seeing instant replays, which we know to be re-presentations of prior events. If you have been asleep since the days of entirely live television, you might take a live picture and its replay to be pictures of merely qualitatively similar events. Thus, we might become used to thinking of echoes as primary sounds heard again with distortions of place, time, and qualities. This capacity is much more fundamental to object perception (see, for example, Scholl and Pylyshyn 1999 and Xu 1999).

Since echo phenomenology arises when the very same event is heard to be at a later time and different place, precisely what we should expect is that the echo should seem distinct from the primary sound. The perceived distinctness of echoes from primary sounds is predicted by the event view.

The event view of sounds therefore has the resources to identify the objects of echo experiences, despite the absence of its characteristic disturbance event from the perceived location of the echo. The account relies on securing the correct way to conceive of hearing a reflected sound. Upon doing so, we see that the event view has the right event on offer—the primary disturbance event.

The problem of echoes is one that must be faced by any theorist who holds that sounds are located at or near their sources and that sounds do not travel (see O'Callaghan 2007*a*). Whether one takes sounds to be properties of their sources, medium-independent events that take place among objects, or some other sort of stationary particular, similar worries concerning echoes and echo experiences apply. The solution I have proposed is, in fact, available to any such theorist. Distal theories of sounds therefore all have the resources to identify the objects of echo experiences despite the absence of their characteristic properties, vibration events, or particulars from the perceived location of the echo. The apparent echo just is the primary sound property, event, or particular.

9.3 Are Echoes Distinct Sounds?

Given that sounds are stationary relative to their sources, and that echoes seem to be located at a distance from primary sounds, it is natural to say that echoes are distinct from their primary sounds. The account I have provided holds otherwise. What reasons have we to think that primary sounds and echoes are not distinct sounds?

Nudds (2001: 227) has argued that if echoes are distinct from primary sounds, then an echo cannot be distinguished from a qualitatively similar sound produced by a different source. We do distinguish between the firework's echo and a similar bang made by a firecracker tossed out one of the brick building's windows. This is supposed by Nudds not to be possible if echoes are distinct from primary sounds. Notice that 'distinguish' must be stronger than 'perceptually distinguish'. We might in the current sense distinguish the echo from the qualitatively similar sound in respects that could not be perceptually discerned. Given equivalent timing, in fact, both scenarios would lead to equivalent experiences.

This argument fails, so it cannot be used against the distinctness claim. To see why it fails, suppose we accept that echoes *are* distinct sounds located at reflecting surfaces. Insofar as the primary sound, the echo, and the qualitatively similar sound from a different source are all full-blown sounds on this assumption, it is no objection that we cannot distinguish the echo from the qualitatively similar sound simply on the basis of its intrinsic properties. Neither sound has intrinsic properties that confer upon it a particular status—for instance, *primary sound* or *echo*—*qua* sound and provide grounds for distinguishing it from the other. But we could, assuming that echoes are distinct full-fledged sounds, distinguish echoes from mere qualitatively similar sounds by their causal relations. Echoes might be sounds caused in part by sounds and sound-generating events that occur elsewhere. The firework's echo would then be the sound at the brick wall that has in its causal history the sound-generating event and sound that occur when the firework explodes. The mere qualitatively similar sound, however, has as its cause a sound-generating event closer to home, and lacks the history of an echo. The sound of a firecracker that goes off near the brick wall may be mistaken for the echo of the firework's sound. But it lacks the right sort of causal relation to the firework's explosion and its sound.

What is 'the right sort' of causal relation? Suppose the firecracker's sound *is* causally related to the firework's sound, perhaps by a sound detector that activates a detonator. Still maintaining that the primary sound and the echo are distinct sounds, we might say that the echo is a sound that has in its causal history a sound-generating event and sound located elsewhere, where nothing intervenes but sound waves that propagate normally. But suppose my striking a tuning fork causes, by no intervening means other than sound wave propagation, a second tuning fork to begin vibrating and sounding on its own. Does the condition fail to distinguish between echoes and resonant sounds? In the case just described, the second tuning fork's sound is generated by the activities of that tuning fork, even though those activities are caused by sound events that occur elsewhere. But an echo is arguably not generated by the events that occur at its location. An echo is a sound that is entirely generated elsewhere. The reflecting surface makes no contribution to the sound, where this notion is fleshed out in terms of sound energy or a measure on the qualities of loudness and pitch. The resonating object, however, contributes to the sound.

According to the proposal we are considering, when sound waves are reflected, a sound occurs. The sound does not travel; the sound waves travel. The primary sound-generating event causes the echo, and only sound waves intervene. An echo is distinguishable from a merely qualitatively similar sound since it is generated by events that occur elsewhere and has a sound in its causal history. This is how to argue that primary sounds and echoes are distinct.

I do not find this satisfactory. It is a plausible principle that sound-generating events do not produce sounds at a distance. If the foregoing were correct, when sound waves from a sound-generating event encountered a reflecting surface, that sound-generating event would produce a new sound. Such 'generation at a distance' might be explained in terms of sound wave behavior, but without causing some new sound-generating event, why should the interaction between sound waves and a reflective surface give rise to an entirely new sound?

We should not take too seriously our pretheoretical verdicts about the causal structure among sound-generating events and various types of sounds. It might be that a stable structure that acknowledges echoes as distinct sounds can be devised. The strategy is not worth pursuing. Despite the phenomenological distinctness of primary sounds and echoes, there are still four reasons that together suggest that echoes are not independent sounds that reside at reflecting surfaces.

First, awareness of an echo normally furnishes awareness of the event that made the sound. Whether this awareness is direct or indirect derives from whether sounds furnish direct or indirect awareness of such events, not anything special about hearing echoes. Both hearing the sound of the firework's blast and hearing its echo furnish awareness of the explosion; if you heard only the echo, you would still be aware of the explosion. Consider the following unusual phenomenon. Someone is hammering on your house. When you are in the right part of the house, and the wind is blowing in the right direction, you cannot hear the sound of the hammering on the side of the house. Now suppose that you can hear the hammering as reflected off the exterior wall of a house down the street. To hear this reflected sound is still to be aware of the hammering. Consider the somewhat common experience of walking toward an intersection and hearing the sound of a drum that seems to be coming from a building ahead of you. When you arrive at the intersection you hear that the drumming is in fact down the street to your left, and that your prior experience was the result of hearing the sound only as reflected from the building. Though you hear the sound thanks to reflected sound waves, hearing it is still a way of hearing the drumming. Awareness of ordinary events by means of echoes is not a deficient way to be auditorily aware of such events. This suggests that what appears as an echo just is the explosion's sound, not a qualitative duplicate that furnishes false or increasingly indirect awareness of earlier events.

Second, we attribute to reflecting surfaces neither audible qualities nor the capacity to produce sounds. Brick walls and canyons are not disposed to make or have sounds when they reflect sound waves; canyon walls and buildings are never loud nor baritone, earthquakes aside. Such dispositions are attributed to the proper range of what we take to be sound sources. But sounds normally seem to be caused, produced, or generated roughly where they are perceived to be located. When a sound seems to be located where there is no sound source, this is likely the result of a physical-cum-perceptual process by which a sound is heard to be where none exists.

Third, what occurs at the reflecting surface is not the introduction of a wave disturbance into the surrounding medium. An elastic collision between the surface and the medium occurs, causing the direction of wave propagation to change. The reflecting body passively redirects the waves. Since there is no new sound source, it is reasonable to conclude that the echo is not a new sound.

Finally, an analogy with the visual case of seeing an object with a mirror is compelling. Just as there is no distinct object located at the mirror's surface, there is no distinct sound located at the reflecting surface. Illusion of place occurs when objects are seen with mirrors, and, likewise, distortions of place and time occur when reflecting surfaces enable us to hear the sounds that occur in front of them.

These four claims together suggest that echoes are not distinct sounds that occur at reflecting surfaces. The account I have proposed explains the apparent distinctness of primary sounds and echoes while maintaining that primary sound and echo experiences share a common external object.

9.4 Are Echoes Images?

It has been suggested to me that an echo is an *image*. There are advantages to thinking that an echo is an image of a sound, but not a genuine sound. This view appears to minimize the illusion we attribute to perceivers. One might veridically perceive an image and its audible qualities near the brick wall, and suffer no more temporal illusion than in ordinary hearing. Reflection might be the occasion for the production of an image of a sound, where the echo is the image. Why think that hearing an echo is hearing a primary sound with illusions of place and time, and not that an echo is a primary sound's image?

We sometimes speak as if images are mental. The image of a musician playing fiddle might stay with you through an afternoon after the musician has gone. Even an apparent echo's existence, however, is not in this way merely mental. Hearing as of an echo may involve illusion, but illusion is *mis*perception. Hallucination is mere seeming to perceive. Even if echo experience involves mishearing, it is not seeming but failing to hear. Echoes are not mere mental images.

It is natural to say that to see or hear an image is to perceive something extramental—it is to perceive a *likeness* of an object or sound. An echo then would be a likeness of a sound. Hearing an echo, however, is not like seeing a photographic image of a face or hearing a 'phonographic image' of a voice. The likeness of mirrors and echoes is not an actual photographic object or phonographic sound that bears likeness to the original. It is not a separate object or sound at all. A mirror image, if it is a likeness, is at most an arrangement of colors at the mirror. An echo then is at most a complex of audible qualities in the absence of a sound that bears them. Thinking of echoes as images of this sort has the advertised advantages.

We need not, however, commit ourselves to such property instances or to the entities we are supposing they constitute: images or likenesses. There is little reason to judge that color instances arranged image-wise exist at the mirror. The apparent colors are fleeting, as are those of a wobbling compact disc. Images so conceived are not required to explain the experience once we recognize (in the auditory case) the primary sound, its waves, and how those waves cause experiences. Just as there need not be elephants or colors at a location in order for you to enjoy an experience that purports to be of elephants or colors at that location, there need not be audible qualities at a brick wall for you to think there are. We do not in general require entities or property instances at a place to account for why an experience seems to be of such entities or properties. That would be to adopt an overly charitable stance on experience, and in the case of property experiences would banish illusions entirely. In the case of reflection, it would multiply property instances at the whim of barriers and experiences.

Since in the auditory example of echoes there is no sound at the reflecting surface to bear the audible qualities, we have little reason to suppose that audible qualities exist at that surface. Optical or acoustic effects in this case suffice to explain the experience and its nonveridicality. If images are complexes of audible qualities, it is therefore unwarranted to posit images at the apparent location of an echo without further reason to believe that a sound exists at that location. We are left with the economical explanation of echo experience as hearing a sound with illusion.

Suppose hearing an image of a sound is enjoying an experience as of a sound that is located where no sound exists. Suppose further that the experience has as its cause a genuine sound located elsewhere. Hearing an image, in this sense, does not commit us to saying that the image is merely mental, or that it has some extramental existence, because hearing or seeing an image is neutral on the nature of the image. This is just to redescribe what I have called hearing a reflected sound or hearing an echo, where a reflected sound is one experienced with illusion due to sound wave behavior.

9.5 Is the Illusion Tolerable?

The event theorist's account of echo perception posits *spatial*, *temporal*, and possibly *qualitative* illusions. Why, since I took the wave understanding to task for its systematic location illusion, is this not objectionable? One might even think that the wave view fares better in the case of echoes, since it posits only a location illusion and an explicable duration illusion.

Illusions, per se, are not reasons for objection. But we are interested in perception for what it can potentially reveal to us about the world. To the extent that we take perception to be a reliable guide to certain aspects of the world, we have an interest in reducing the amount of illusion we attribute to perceptual experience. But illusions can inform us about the mechanisms involved in perception and about aspects of the world that we cannot directly observe. So we also have an interest in discovering the illusions we actually fall prey to. If, however, an account proposes an illusion whose spell we have independent reason to believe we are not under, all else equal we should prefer an alternative that does not posit that illusion.

I have suggested that we have no reason to think that we are under the illusions posited by a wave theory of sounds, and also that in the case of echoes there are compelling reasons to believe that we do fall prey to the illusions posited by the event view. The event view's illusions therefore have a significantly different status from the wave view's.

According to the account I have given of echoes and echo perception, the illusions arise as a *special case*. While in ordinary sound experience we hear sounds for the most part as they are, the experience of an echo occurs only in quite special circumstances. The illusions posited by the wave conception are systematic and pervasive.

The event theorist's illusions are *explicable*. Quality illusions occur because filtering and scattering that occur at reflecting surfaces distort information contained in sound waves. Illusions of place and time occur because sound wave reflection mimics the situation in which a sound source exists at the reflection site. So there are external states of affairs that explain why experience attributes the properties it does when the illusion occurs. Because of this, the illusions in the case of echo perception are also *predictable* once the mechanisms of ordinary sound perception are known.

Finally, the illusions that occur during echo experience are analogous to visual illusions that we find interesting but unproblematic.

Hearing the firework's echo to be at the brick wall is like facing a large mirror and thinking there is a piano in a room ahead of you. Hearing the echo to occur after the primary sound is like seeing a supernova from across the galaxy.

You might worry that if the event view is correct, the time gap that occurs even in ordinary sound perception—because sound waves travel relatively slowly—is problematic. But it is simply an exaggeration of the time gap that occurs in vision and goes undetected unless somewhat large distances are involved. According to the view I am defending, David Armstrong was incorrect to think that sounds might *not* pose a time gap problem.

> In the case of a star, it may be questioned whether our immediate perception really involves any temporal illusion. It may be suggested that what we immediately perceive is not the star, but a *present* happening, causally connected with the extinction of the star many years ago. The star sends a *message* to us, as it were, and we immediately perceive the message, not the star. Now this suggestion may be correct in the case of a sound. There seems to be some force in thinking of sound as actually spreading out from its source, like a balloon rapidly inflating. (And here I am not speaking of the sound-*waves*.) So when two people 'hear the same sound' it may be argued with some plausibility that they *immediately* hear two different things, because they are in different positions. (Armstrong 1961: 147–8)

Armstrong goes on to reject the analogy in the case of a star, since the only immediate object of sight is 'the star itself' and not light waves or energy. He concludes that sometimes with vision we immediately perceive past happenings, though they seem present. I find it implausible that a sound is like a rapidly inflating balloon and that two people cannot hear the same sound. So I reject Armstrong's characterization of sound and accept that hearing past sounds is like seeing past events. Both involve a time gap and a temporal illusion.

I have provided an account of echoes and echo perception that complements the view that sounds are distal medium-disturbing events. According to this account, an echo just is a primary sound whose perception involves distortion of place, time, and perhaps qualities. Echoes are not distinct from primary sounds. The impression of distinctness occurs because when we immediately perceive past happenings we perceive them as present, and because without special tricks we perceive events in their entirety only once.

Once we recognize that echoes are not distinct from their primary sounds, we need not be forced into the view that sounds are object-like particulars that persist and travel. That view rests upon shaky foundations and cannot accommodate the distinctive way in which sounds inhabit time. The argument from echoes fails to establish the conclusion that sounds are not events because echo experiences involve reencountering a sound event thanks to a special trick. Hearing thanks to waves that rebound from a reflecting surface is just a way of hearing a sound again. Echoes pose no novel problem for the theory that sounds are events. We should therefore prefer the event view of sound, according to which a sound is the event of an object or interacting bodies disturbing a surrounding medium in a wavelike manner.

10

Hearing Recorded Sounds

10.1 The Puzzle of Recorded Sounds

When standing in a room, engaged in conversation, you hear the sound of your conversant's voice. Hearing the sound of spoken words fosters awareness of the speaker and the voice. Hearing the sound of a mandolin or of glass breaking, likewise, furnishes perceptual awareness of objects, activities, and events in your environment responsible for those sounds.

Suppose, however, that you are sitting at the back of a lecture hall out of earshot from the lecturer and that the sound you hear is produced not by a person but by a loudspeaker mounted on the wall behind you. You hear a sound produced by the loudspeaker, but you also seem to hear the lecturer's voice. If, however, the sound you hear is clearly that of the loudspeaker, have you heard the lecture at all, or have you heard a mere representation or facsimile of a lecture cleverly produced by the technology of loudspeakers?

The natural assumption is that you indeed hear the lecture—why else stick around? But this case falls on a continuum, the far end of which makes similar judgments difficult to reach. Take one step: you hear a radio broadcast of the president's speech. What you most immediately hear is the sound of your radio's speaker disturbing the surrounding medium, but it is again natural to suppose that you have heard the president and his latest address. Suppose the speech is delayed a moment by wire transmission between Washington DC and New York, or several seconds from around the world. Now, suppose it was recorded earlier in the day. Does listening to a delayed relay or a recording amount to an episode of perceptual awareness of a far-off sound and the events that produced it? Finally, imagine that you visit the Library of Congress and fetch a recording of an old FDR speech. Hearing a reproduction of FDR's voice over headphones in a dark archive intuitively lacks the red-blooded markings of genuine auditory perceptual acquaintance with the former president himself.

The puzzle for the theory of sounds and their perception is that, on this continuum, verdicts about whether perceptual acquaintance occurs differ despite the structural similarity of the cases. In each example, information about the original sound event is through technical means transduced, preserved, transmitted, and reproduced as a new sound. The original sound in these cases causes the subject's auditory experience, and the experience depends upon the sound in a way that supports the relevant counterfactuals about how the subject's experience would have been if the original sound had been different. If this causal, counterfactual supporting dependence is a necessary condition on perception, if the condition is met in these kinds of cases, and if subjects' experiences are true to the original sounds, then either meeting some further condition or constraint raises causally and counterfactually dependent awareness to perception in some of the cases or some of our differing judgments across the range of cases are mistaken.

On the end of the continuum of cases nearer to ordinary unassisted perception are those about which it would be overly austere to deny real perceptual awareness. Suppose you hear my voice during our conversation thanks to your set of the latest hearing aids. The hearing aids produce the amplified sound in virtue of which you listen to my voice. Suppose now that the hearing aid transduces the available wave information into electrical impulses and directly stimulates your cochlea (as with cochlear implants) or auditory nerve. To deny that this amounts to hearing is to commit oneself to an arbitrarily strict understanding of auditory perception. Similar prosthetic examples, though technologically more far-fetched, can be constructed for vision.

Though sound reproduction provides a compelling instance of the puzzle, the issue is not peculiar to audition. Live video, film, and even photographs enlist technologies that ground information-preserving causal connections between ordinary objects or events and visual experiences. Few may deny that you see the space shuttle Discovery and its liftoff during a live television broadcast, but Kendall Walton's (1984) claim that photographs are transparent—in that they foster genuine seeing of what they depict—has met strong resistance. The claim that seeing a photograph is a way of seeing FDR has been understood to count against any account of perception from which it follows.

Sound reproduction might, nonetheless, pose unique trouble for the event conception of sounds. As I have developed it, the event view holds that the transmission or reflection of sound waves by barriers fosters an experience of the originating sound event. Unlike the case of echoes or transmission through barriers, however, it is uncontroversial, even according to the event theory, that radios and telephones generate new sounds. All the same, information about the originating sound event is preserved by both reflection and telephones. Since, for the event theorist, waves simply are the customary means by which information about sounds is transmitted, event theorists might be under particular pressure to accept that telephones and even recordings precipitate hearing the original sound.

Resolving the puzzle for auditory perception involves addressing two sets of questions. The first deals with the conditions required for perception, in general. In short, does interrupting or altering the natural causal process that ordinarily grounds awareness of objects and events rule out perceptual awareness of those objects and events?

The second set of questions deals with the distinctive characteristics of auditory perception. Though sounds are the proper and immediate objects of auditory perception, audition also reveals to us a world of happenings, activities, and their participants. How might auditory access to the world of things and events differ from vision in respects that must be acknowledged by a full account of hearing transmitted or recorded sounds? Could the differences between the objects or contents of audition and vision ground a difference between perceptual access in one modality and no such access in the other?

I argue in this chapter that indeed there is a sense in which you hear not only the sound of the loudspeaker or headphones, but also the original sound that was recorded or transmitted. Sound reproduction technology grounds a form of perceptual access to the sound of FDR's voice. FDR himself thereby is perceptually available to audition. This access, however, is *mediated* and, therefore, *indirect* in a sense I shall explain. This form of indirectness, in contrast to some other forms discussed in the philosophical literature on perception, does not preclude a kind of genuine awareness. Your mediated awareness of FDR's voice and his address, furthermore, yields an experience that is *illusory* and *impoverished* in critical spatial and temporal respects. No barrier exists, however, to enjoying a variety of perceptual acquaintance with the former president that reveals the sounds of his words and their defining qualitative characteristics.

10.2 Spatial Perspective in Perception

What makes being at the game different from seeing the game on television? The standard response of causal theorists about perception has been that the *normal* or *natural* causal process involved in ordinary unassisted perception fails to obtain and that such a causal relation is a requirement for perception (see Strawson 1974; compare Grice 1961). The problem with this response is not that it appeals to an empirical characterization of the natural causal process involved in unassisted perception, but that it is far too strong (Noë 2003). Virtual reality glasses and surgically implanted vision aids of the future, as well as present-day hearing aids and cochlear implants, qualify as making possible genuine perceptual experience. None, however, preserves the normal causal process.

Ordinary visual perception, however, not only presents its objects as being a certain way—as having certain qualities like color and spatial properties such as shape, volume, and location—but also presents characteristics that depend upon your perspective on the environment. When you see the barn, you see it not only as having a certain color, shape, and size, but also as occupying a certain location and as having an apparent shape relative to where you are standing. You might also see certain highlights and color variations the barn appears to have from your present vantage point, or see it to occupy more of your field of vision than the adjacent house, whether or not you see it to be actually larger. Gilbert Harman explains the perceptual experience of such relational features like this:

> Various features that the tree is presented as having are presented as relations between the viewer and the tree, for example, features the tree has from here. The tree is presented as 'in front of' and 'hiding' certain other trees. It is presented as fuller 'on the right'. It is presented as the same size 'from here' as a closer smaller tree, which is not to say that it really looks the same in size, only that it is presented as subtending roughly the same angle from here as the smaller tree. To be presented as the same in size from here is not to be presented as the same in size, period... I mean only that this feature of a tree from here is an objective feature of the tree in relation to here, a feature to which perceivers are sensitive and which their visual experience can somehow represent things as having from here. (Harman 1990: 38)

Perspectival content, including relative or egocentric location, varies not only with changes in the environment, but also with changes to the subject since it is relational content. If the barn or the sun highlighting it moves, perception reveals the barn as changing its position, apparent shape, or pattern of illumination relative to you. If you move about the barn or lie horizontally in the grass, your perspectival relationship to the barn changes, as does your experience of that relationship: lying down, the barn seems turned ninety degrees relative to your field of view.

The perspectival characteristics of a visual experience, in addition to its qualitative and non-perspectival characteristics, ordinarily depend counterfactually on the perspectival relations between a subject and the scene. Noë (2003) and Cohen and Meskin (2004) diagnose the problem with the appeal to natural causal processes by suggesting that such perspectival dependence is necessary for an experience to count as fully perceptual. According to Noë and to Cohen and Meskin, the standard causal account of perception thus suffers not from a conception of the relevant causation that is too weak, but from an understanding of the contents of visual perceptual experience that is too thin.

When, for example, a subject views a televised image of a speech, the perspectival characteristics of the experience do not depend in the right counterfactual supporting way upon the spatial relationship of the viewer to the speaker. The president could move from side to side while occupying the same relative position on the television screen. You can move from the living room to the kitchen and view nearly the same image of the speech. The televised image does not change in the subtle ways that perceptual content does as you move your head and eyes to explore a scene. You cannot, for instance, see the president's ear by moving around to the side of the television. The televisual process does not preserve information about the spatio-temporal relationship between viewer and events depicted.

The counterfactually dependent content of the visual experience thus falls well short of that of perceptual experience of objects and events that are in one's environment. The accounts both of Noë (2003) and of Cohen and Meskin (2004) entail that film, video, television, simulcasts, photos, and the like all make possible experiences whose qualitative and non-perspectival (*factual*, according to Noë) characteristics are causally connected with and counterfactually dependent upon those of the objects they depict. Cohen and Meskin (2004) put this in terms of the information that photographs make available, which is information about the *visually accessible* properties of the object, but not about the *egocentric location* of the object. Experiences initiated by these visual aids do not, however, depend counterfactually upon the perspectival relationship between a subject and what the recorded or transmitted image depicts. This lack of counterfactual dependence between perspectival properties as presented in experience and those in the world that hold in virtue of the relationship between subject and scene means that the corresponding experiences cannot be fully perceptual. What makes ordinary perceptual experience distinctive is its thoroughly perspectival character; a counterfactual supporting causal dependence of characteristics of experience upon characteristics of the world is a necessary condition on perception; so, a condition on perception is not met when seeing photos, films, videos, or simulcasts. The condition, however, is met when visual experience is causally coupled to worldly perspectival relations as it is when, for example, a subject wears special video goggles or receives perfect artificial stimulation of the optic nerve.

10.3 Perspective in Audition

Auditory perception, too, is perspectival. One experiences and learns through audition where the chirping bird is from here (above, ahead, and to the right), and auditorily discerns that a train is approaching and passing. Though audition is neither as accurate nor as precise as vision, it provides an important mode of access both to the spatial characteristics of an environment and to where one stands in that environment. The perspectival character of auditory perception changes as one's relationship to the sounding objects and events in one's environment changes. Moving closer to a source, for example, increases its apparent loudness relative to other sounds though one perceives it to have the same volume. Relative motion shifts apparent pitch, though one need not perceive the sound's pitch to have changed.

If the above 'missing perspectives' account of why watching a film is not seeing its subject is correct, then a similar story should hold for hearing a recorded sound. According to this account, when you sit at the back of a lecture hall hearing a loudspeaker's sound, or listen to the radio, you fail to hear the sound of the lecturer's, or announcer's, speech. Since the spatial perspectival properties of the original auditory scene do not covary with those you experience when listening to a recording or a live broadcast, the account should run, you fail to hear the particular sounds and events in that scene.

The trouble is not that this account fails to discern an important difference between ordinary sound perception and hearing recorded or transmitted sounds. The trouble is also not that there exists a fundamental difference between audition and vision that makes the account untenable in the case of hearing. It is, perhaps, more compelling in the case of audition than it is in vision to think that you perceive the original. Intuition at least suggests more strongly that perception survives transmission in audition: we say with confidence that we hear our conversant's voice during a telephone conversation, but more readily abandon thinking that we really see the president during a live televised speech. I discuss in the next section whether there could be such a perceptual difference and diagnose the source of the difference in intuitions. I want to point out now, however, that despite the intuitive difference, the cases appear structurally similar, and so are prima facie on a par with respect to whether perception takes place.

The issue with the missing perspectives account, I want to suggest, is that although it successfully distinguishes ordinary or otherwise fully veridical perception from experience generated through transmissions or recordings, it fails to capture the real and convincing sense in which we are aware of the original objects or sources of such experiences—the sense in which you see the president on television or hear the sound of the lecture.

To be fair, Noë (2003: 95–6) acknowledges that there is an important sense in which you see the game on television, and that to deny this is dogmatic. However, he is interested to spell out what constitutes the difference between televisual seeing and genuine or ordinary veridical perception. The understanding of perception as essentially involving rich veridical perspectival content that supports corresponding counterfactuals is his contribution to characterizing what he calls perception *simpliciter*. But if that rich content is necessary for perception *simpliciter*, what is left of the sense in which we see or hear the original? My interest in this chapter is whether and how it can be correct to say that one hears the lecture over the loudspeaker given observations about the perspectival content of perception. Considering why perspective matters reveals why telephonic auditory experiences are a form of perceptual acquaintance.

I begin by characterizing the seeing and hearing that occurs in real-time cases and extend this explanation to cases of genuine seeing and hearing when the objects, events, and sounds perceived no longer exist. It is precisely the sense in which photographs and recordings ground a variety of perceptual access to the originals that I wish to articulate.

10.4 Do Kind Differences Matter?

Intuitively, whether or not it is reasonable to believe you fail to see the president when you watch a televised address, it may seem that denying that you hear the sound of his voice during a broadcast is an overly strict consequence of perceptual theorizing. Do apparently analogous auditory and visual cases differ in ways that make the difference between perceiving and not perceiving? If so, the intuitive difference may diagnose a genuine one. If not, our account of technically mediated awareness should be responsive to the data from audition as well as from vision, and we should discern the source of the intuitive difference between the two classes of cases.

What factor could make the difference between perceiving in the real-time auditory scenario and not perceiving in an otherwise comparable visual scenario? The most obvious candidate is that, in the case of relayed sound, in contrast to television, the object of immediate experience is of the same metaphysical kind as the purported object of mediated awareness. The sound you hear over the loudspeaker is the very same general sort of thing as the sound generated at the original source. A televised image, however, is not a person or an event involving things of the same broad kind as those depicted (unless you view an image of a television image). Audition, unlike vision, guarantees experience of the very same kind of particular—a sound—when hearing a transmitted or replayed signal. It would be an incredible technology that could produce a president in your living room for your family to watch deliver a speech.

Despite the significance of this difference, it reveals no barrier to perception exclusive to the visual case. The metaphysical kind of that in virtue of which you enjoy awareness should make little difference to whether your awareness counts as awareness of a distinct particular, even if the two differ in kind. If the experience seems to be as of seeing the president waving his hands, and if the experience is made possible thanks to an appropriately caused televised image to which you attend, then the kind difference between images and persons should not by itself preclude perception. It is more plausible that seeing occurs with the help of video or night-vision goggles that preserve perspectival content, so verdicts about whether perception occurs are not driven just by the mediator's kind. Whether that which mediates awareness matches in kind what one seems to experience is not the deciding factor. More important is the distinctness of the mediator from the object of which one seems to be aware, along with the fact that such awareness does not require inference or further cognition. Consider the possibility of the technology mentioned above: that of producing a replica president in your living room to deliver the speech. If seeing an image of the president on your living room set is not perceiving the president, then seeing a replica of the president in your living room to deliver the same speech is not perceiving the president. Sameness of metaphysical kind between the mediator and the original is insufficient to make the difference between perceiving in the auditory case and not perceiving in the visual case.

The intuitive difference between the auditory and visual cases, however, stems from a connected point. Our interest in sounds often is due to the informational value they have in virtue of being of a certain qualitative type. Spoken words, for example, instantiate complex sound types characterized by specific patterns of qualitative change though time. The sound of the spoken word 'lever' differs from the sound of the word 'sever' precisely because it involves a different pattern of audible qualities over a similar interval. The sounds of birds' calls differ in comparable ways and make it possible to identify particular species by identifying the qualitative types of their characteristic sounds. Musical sounds, which I discuss in more detail in the final section, instantiate complex sound types that arguably are the best candidates for musical works.

Sound reproduction not only guarantees unmediated awareness of a particular of the same metaphysical kind as the original, but also produces a new sound particular of the same qualitative audible type as the original when the reproduction has good fidelity. Experiencing the reproduction thus has roughly the same informational value as the original. This contrasts with visual reproductions, which, though they preserve much qualitative information, do not generally belong to the same characteristic identifying types as the originals. The changing pattern of colors and shapes that constitutes a television image does not itself instantiate the type that grounds our interest in persons and soccer games. There are exceptions in the visual case that support the general point about audition: high-quality photographs of photographs or other works of art; televised images of signed language; scanned PDF images of documents viewed on a computer screen. Each of these instantiates the same relevant qualitative type as the original, but in these cases the originals themselves are of interest particularly in light of the informational value of their qualitative types. This focus on the qualitative types of sounds driven by a strongly information-driven interest in sounds may lead to the difference of intuitive confidence between judgments about whether perception takes place when hearing a voice over the telephone and when seeing a televised image of a person or event. Transmitted or recorded sounds thus are informationally less impoverished than visual images. The significance of this difference in degree is, I suggest, confined to intuition and does not constitute a difference between perceiving and failing to perceive.

10.5 Perspective and Perceiving

Televisual and telephonic experiences thus are on a par concerning whether some form of perception takes place. In what sense, if any, do both kinds of cases involve a variety of genuine perception?

The first apparent obstacle is that both involve at best *mediated* awareness. If you hear the sound of your conversant speaking during a telephone conversation, you hear it by hearing the sound of the telephone's speaker. If you see the president when watching television, you see him by seeing a televised image. We require an account of the possibility of mediated perceptual awareness. The relevant sense of 'mediated awareness' is more innocuous than the sense involved in discussions of 'indirect' and 'representational' accounts of perception according to which a mental intermediary or representation mediates awareness of all external perceptual objects. In the sense relevant here, awareness of some object, event, or sound is mediated awareness when one enjoys an appropriately caused experience as of that thing in virtue of perceptual awareness as of some distinct object, event, or sound. The experience as of the sound of the lecture, for example, occurs thanks to your awareness of the sound of the loudspeaker. Though you hear the loudspeaker's sound with no intervening object of awareness, you hear the sound of the original lecture thanks to your awareness of the loudspeaker's sound. You might attend to the sound of the loudspeaker and discern its qualities as an independent object of experience, but no awareness of the sound of the lecture occurs without hearing the loudspeaker's sound. An individual wearing video goggles or hearing aids might, for example, attend to the video image or hearing aids' sounds as such, though still be capable of mediated awareness of the surrounding environment with the aid of these prosthetic devices. In short, this variety of mediated perceptual awareness requires that one enjoy an experience as of the original particular, that one enjoy an experience as of the mediator, that one's awareness as of the original occurs in virtue of one's awareness of the mediator, and that one's awareness as of the original particular depends causally and in a counterfactual supporting manner upon that original particular.

What is striking is that awareness of a loudspeaker or of a television screen should afford experience as of a voice or a person. As Noël Carroll (1985) points out, however, the capacity to recognize a person suffices for the capacity to recognize that a televised picture depicts a person. Analogously, the capacity to recognize a person's voice suffices for the capacity to recognize that a loudspeaker's sound projects a person's voice. The experience of pictorial images and transmitted sounds thus differs phenomenologically from the experience of convention-based representations. The experience makes it possible to recognize the picture's subject or the sound the loudspeaker relays. In light of this kind of convention-ignorant recognition, awareness of a picture or of a loudspeaker's sound affords a variety of experience as of a person or a voice. Unlike conventional representations, therefore, an intervening image or sound need not pose a barrier to a variety of perceptual access. On the contrary, explaining recognitional capacities afforded by such images and sounds might demand it. Though mediated awareness is parasitic on at least potential attentiveness to the mediator, it does not rule out a kind of genuine perceptual awareness.

I can step on your toe, even though both of our shoes mediate the transaction. Likewise, I often talk to my wife on the phone. The connection we make is certainly 'mediated' in all sorts of ways; still, it is my wife that I talk to, and it is her voice that I hear. (Fodor 2000)

The kind of mediated awareness involved in hearing amplified sounds or in viewing televised images, however, lacks the accurate perspectival information about, for example, egocentric location that Noë and Cohen and Meskin identify. In that respect, this kind of mediated awareness is impoverished along one dimension of its content. You cannot find out about the location of the lecturer relative to you just by hearing his voice with the help of the loudspeaker behind you. Why, therefore, does mediated awareness count as perceptual awareness at all? Given Noë's and Cohen and Meskin's claim that veridical perspectival content is necessary for perception, and given that such information may play a crucial role in determining just what counts as the object of perceptual awareness, one might object that mediated awareness as I have characterized it is mere metaphorical hearing or seeing. Perspectival content certainly is critical to characterizing an experience as perceptual: arguably, experiencing without perspective is not perceiving. Systematic patterns of variation to perspectival content are critical requisites to experiencing something as a particular distinct from oneself (see, in particular, Strawson 1959, Evans 1980, Noë 2004, Siegel 2006*b*). To hear a sound as independent from oneself—to perceive it as an objective particular, in contrast to merely enjoying auditory experience—involves implicitly grasping how aspects of one's experience and properties of the sound itself do and do not depend upon and vary with changes to one's perspective.

Impoverishment of veridical perspectival content, however, does not imply that the experience of a recorded sound contains no perspectival content. In fact, the experience presents its mediated object as bearing a perspectival relation to the subject that is determined by the mediating object. So, for instance, the experience appears to present the sound of the lecturer's voice as coming from the direction of the loudspeaker behind you.

And though we are rarely taken in, watching the president on television seems experientially to locate him where you see his image to be from the camera's point of view. It seems to present him as standing in a certain relation to the perspective of the camera. We adopt the perspective of the camera in televisual seeing (see also Noë 2003: 95). You reflexively duck when the projectile seems to fly straight toward you at the movies because you reflexively accept the camera's perspective as your own.

That perspective, however, is illusory for you. Consider seeing a mirror image of the president. The visual experience of the mirror image seems to present the reflected individual as being located in the direction of sight. Hearing a good sound recording presents information about the locations of the musicians relative to the microphone. Though experienced perspectival content varies with changes to the perspectival relationship between the camera or microphone and the original object, event, or sound, it misrepresents or fails to reveal your actual perspectival relationship to the source. Our resistance to accepting such experiences at face value is learned to the extent that we have learned to distinguish the mirror's, camera's, or microphone's perspective from our own. However compelling, the experience in real time of a transmitted sound or image is perspectivally illusory because the subject does not in fact occupy the point of view of the microphone or camera.

Despite its impoverished veridical content, an experience of a transmitted sound or image constitutes a form of perceptual contact with the original source in virtue of our adopting a point of view that we do not in fact occupy. Since the experience invokes the perspective of the camera or microphone, that content suffices, along with the veridical qualitative and non-perspectival content, to grasp the original source as an autonomous object of perceptual acquaintance. It suffices, that is, to discern the original source as an independent particular distinct from its visible or audible image.

Mediated awareness therefore bears the marks of a variety of contact. One's contact is of course mediated by the sound of a loudspeaker, to which one can attend by focusing upon one's actual perspective on the local auditory scene. Nonetheless, one can have an experience as of the sound of a lecture taking place or as of seeing a soccer game thanks to the experience of a loudspeaker's sound or a television's image. The reliability of the technology ensures the experience is appropriately caused and depends counterfactually on the lecture or the game; the experience has veridical factual or non-perspectival content; and the experience includes perspectival content, which, though illusory, purports to situate one with respect to the original sources. This is the sense in which you are aware of the sound of the original lecture when listening to the speakers at the back of the hall, and the sense in which you see the game with television.

Unless there is some reason to believe that *veridical* perspectival content is a necessary condition on genuine perception of a particular, these cases are best characterized as involving illusion and not hallucination. Illusion, after all, is genuine awareness that falls short of entirely veridical content in some respect. To misperceive the size, shape, or color of the tree from here is to perceive the tree, but it is to experience it as having some property that it lacks or that is unconnected to your current experience as of its having that property. To hallucinate, on the other hand, is to have an experience that seems to be of a particular object but which is not perceptual. Vivid dreams are hallucinations; seeing a Necker cube as three-dimensional is an illusion. Illusion is misperception, but hallucination is a failure to perceive. Seeing with mirrors and hearing echoes and refracted sounds demonstrates that perspectival content might be illusory without experience collapsing to hallucination. Hearing a speech over radio is a way to enjoy an experience as of hearing the sound of the speaker's voice. With ordinary radio, our experiences are causally connected with and counterfactually dependent upon the sounds of the original sources of the audible images. Listening with radio, therefore, is not a variety of hallucination.

Noë and Cohen and Meskin are right that ordinary experiences frequently include veridical perspectival content, but ordinary and not-so-ordinary cases of perception may also be deficient in veridical perspectival content. Seeing with mirrors, periscopes, microscopes (including optical and, for instance, scanning electron microscopes), telescopes, and bifocals involves illusions of perspective on the objects experienced. So does hearing with ear trumpets, parabolic sound disks, two cans attached by string, reflecting surfaces, and other acoustic appliances. Seeing in low light and hearing in noisy environments also degrade perspectival content. It would be dogmatic to deny that perceptual connections exist in these cases.

Perceiving is a matter of degree. Experiences can be more or less veridical, more or less illusory. The content of a perceptual experience can be factually, qualitatively, and perspectivally quite rich, or it can be highly impoverished. In an important sense, radio grounds the possibility of genuinely perceptual experiences whose perspectival content falls short of the rich and veridical content of everyday, unassisted auditory experiences of items within one's local environment. Impoverishment of perspectival content need not make the difference between perceiving and not perceiving, but as perspectival content gets thinner and less accurate, the intuition that the experience counts as perceptual weakens. And it should. Without some measure of perspectival content, it is difficult to understand the experience as being sufficiently determinate to pick out an object in the surrounding environment (especially when that environment is the camera's or microphone's). Such experiences threaten to collapse into mere sensation. Perception with the aid of radio and television is possible, but such perception is, in the senses discussed above, mediated and illusory. It is, therefore, impoverished of veridical perspectival content, but it is not impoverished entirely of perspectival content.

10.6 Hearing and Seeing the Past

The discussion so far has focused on assisted perception of objects and events in real time. But the account extends to experiences produced by listening to replays of events and sounds recorded earlier. That is, hearing a replay of a speech recorded earlier in the day, or twenty years ago, seems also to count as a form of impoverished or illusory perception. How, one might object, is it possible to have genuine perceptual access to something that no longer exists? If anything counts against the claim that a subject enjoys perceptual access to some object or event, the non-existence of that object or event should. FDR no longer exists to be heard and JFK no longer exists to be seen, even with the help of recordings, films, and photographs.

I want to suggest that even recordings, films, and photographs ground a variety of perception, though, again, it is mediated, impoverished, and illusory.

Ordinary perceptual experience includes, in addition to spatial perspective, a *temporal* perspective. The objects and events one experiences perceptually seem not just to exist at a particular time, but to occur *now*. The experience of a soccer game or of a concert presents it as taking place at the present, as at the same time one experiences it. And experience does not, in ordinary cases, present things whose relationship to the subject is not (near enough) that of concurrence. One might have the perceptual sense that the sound of the note one hears at present has continued for some time, and one might have the perceptual expectation that it will end in a certain way. Nonetheless, one cannot perceptually experience something to have occurred long ago. One can, however, succumb to an illusion that something long past occurs now. Seeing a star is seeing something long past, though one's experience includes the illusion that it now exists. Seeing Mars in the night sky or seeing a ship in the distance is seeing something as it was, though the illusion of temporal perspective is less severe.

When events change their temporal relationship to your current outlook—for example, by ending—your experience ordinarily ceases to present them as concurrent with your present temporal outlook. Though temporal perspective may include, in addition to *presence*, a horizon of immediate *past* or *future*, our perceptual repertoire of temporal perspectival properties is substantially less than that for spatial perspectival properties. (Imagine only being able to perceive something as *there* when it is ten feet in front of you at eye level; perhaps this is like walking in pitch dark with a flashlight pointed at the ground in front of you.) This probably explains the intuitive appeal of presentism. If things and times other than those present exist, then presentism's mistake is confusing an account of the contents of perceptual experience for an account of reality. Presentism mistakes a failure to be perceived for non-existence. God, however, is guaranteed a much more extensive repertoire for temporal perspectival properties, and probably is not a presentist.

Noë's and Cohen and Meskin's analyses thus can be applied to show another dimension along which recorded images fail to ground fully veridical perceptual experiences of what they depict. Films and recorded video do not cause experiences whose temporal perspectival properties depend on the actual temporal relationships between the original objects and events and the subject's present temporal outlook. You can watch a taped speech over and over with no experiential (temporal perspectival) distinction between it and a live simulcast. Each seems as if it takes place *now*. And though a holographic projection caused by and visually indistinguishable from FDR may fool you into thinking the man himself is now before you, you suffer an illusion of temporal perspective. In each case, adopting the temporal perspective of the camera or microphone explains the illusion.

Hearing a recorded sound thus is a way of listening into the past. You hear the sounds and events that occurred at the time of the recording, but experience presents those sounds and events as taking place during the time of listening.

How photographs differ from recorded sounds and filmed or videotaped scenes is precisely what grounds their interest and value. A photograph cannot by itself reveal change. The technology results in an image that remains static through time, though the objects and events it depicts were changing when the image was captured. Photographs, therefore, are impoverished in yet another perspectival respect: they present an unshifting temporal point of view on a given moment. This perspective, unlike the temporal perspective of a perceiving subject, is incapable of revealing change precisely because it reveals things as they are only at a moment.

But photographs also highlight, by contrast, what I have been emphasizing since the start. An audible analog to a photo—a sound frozen in time—has very little interest apart from special situations and needs. Sounds hold our concern in light of their behavior through time. Technology to capture a sound without change or duration misses precisely what grounds the allure and value in sounds.

10.7 Hearing Musical Performances

Listening to broadcast or recorded music is an important instance of the general puzzle. When listening to your stereo's sounds, have you heard Hendrix's rendition of 'The Star-Spangled Banner'? If not, what have you heard? But if listening to a recording does amount to hearing Hendrix's performance, what accounts for the aesthetically significant difference between hearing a recording and hearing a live performance?

According to my 'illusory perspectives' account, hearing the recording in your living room is a way of perceiving the sound of Hendrix performing 'The Star-Spangled Banner'. Your experience of that sound, however, is impoverished and illusory in a number of respects, including its spatial and temporal perspectival contents. Several important points concerning what you hear and its aesthetic value need attention.

First, though your awareness of the sound of Hendrix's performance is illusory and impoverished along the perspectival dimensions of content, you do enjoy full-fledged veridical awareness of an instance of 'The Star-Spangled Banner'. When you listen to the sound produced by your stereo's speakers, that sound is of the same sound type as the sound of Hendrix's performance. The song itself is a complex sound type constituted by a particular pattern of notes arranged through time. A pattern of sounds counts as an instance of a song just in case it includes sounds of the appropriate pitch and duration, arranged according to the right timings. Recordings typically guarantee just that. Listening to your stereo therefore counts as enjoying the music you set out to hear. There is no question of mediation or illusion on this count.

What, then, accounts for the aesthetically significant difference between hearing a high-fidelity recording and hearing a live performance? Hearing a recording furnishes a limited form of perceptual access to the performance itself mediated by your awareness of the current sound. In addition to being mediated and indirect, in virtue of the process of recording and reproducing the sound, your awareness of the original performance also is both impoverished and illusory along a number of dimensions.

To start, in the case described, your experience of the performance occurs in a single modality. Adding visual information might change the aesthetic value of the experience, but it is difficult to duplicate the multisensory experience of a live performance. So consider the aesthetic characteristics revealed in the auditory experience alone. Even a nearly indistinguishable auditory experience of a recording aesthetically lags the auditory experience of hearing a live performance. Because it lacks the veridical perspectival content of hearing a live performance, hearing a recorded sound falls short of the kind of immediate auditory perceptual access to a performance that one has when attending a live show. Hearing a recording is, most immediately, hearing a sound of the same qualitative type as one produced by a performing artist. It is only in a more distant and impoverished sense the hearing of a *performance*, which is the act of a musician or vocalist generating sounds through skilled conscious manipulation of an instrument or voice. A performance surely has aesthetic value that a sound does not. And, surely, having more direct perceptual access to a performance makes an aesthetic difference. Hearing recorded music in this respect differs aesthetically from hearing a live performance.[1]

[1] McGill University cognitive neuroscientist Daniel Levitin, in recent unpublished research, reported by Elton (2006), has attempted to quantify this difference by measuring physiological and emotional responses both to recordings and to live performances of the Boston Symphony Orchestra.

11

Cross-Modal Illusions

The account of sounds and their perception that I have developed in this book holds that sounds are a certain kind of event which we hear thanks to information borne by sound waves. Sounds, I have argued, are not among the traditional secondary or sensible qualities because sounds are particular individuals. Sounds, however, are not object-like particulars. Though for the purposes of a theory of sounds, pressure waves are best understood as events that occur in the medium, and though waves are the proximal stimulus to audition, waves themselves are not among the objects of auditory perceptual experience. The medium, according to the event view of sounds, is a necessary condition on both the existence and the perceptibility of a sound, but the medium cannot satisfy vital phenomenological and epistemological constraints on the proper objects of auditory perception. The key is that sound waves transmit information about sounds but are not identical with the sounds; the lesson is that sounds occur and have durations we commonly experience. This account relies on a model of auditory perception that differs in important respects from the received understanding according to which hearing just is responsiveness to patterns of pressure difference that propagate through a medium. Successfully hearing a sound, on the account I have proposed, involves more than just an experiential response to sound waves. Distal events constrain what counts as veridical auditory experience.

A particular sound, on the view I have proposed, is the event of an object or interacting bodies disturbing the surrounding medium in a wavelike manner. This proposal captures the fact that sounds are heard to be localized in the environment and, thus, that sounds provide information about the spatial features of one's surroundings. It recognizes that the audible qualities are attributes of sounds that depend upon the physical characteristics of the interaction over time that takes place between the disturbing object and the medium. Pitch, timbre, and loudness provide auditory analogs of color and other sensible qualities.

The discussions of the foregoing chapters establish that the event view furnishes the materials for an explanatorily robust understanding of sounds as objects of auditory perception. Though I have described a number of situations in which subjects misperceive sounds so conceived, these cases prove to be exceptions that strengthen the general account. Interference, echoes, and Doppler effects each present an opportunity to show that a perceptually ambiguous situation should be resolved in favor of a simpler or more elegant hypothesis about the world of sounds than wave-based approaches provide. Where sensory stimulation itself is ambiguous, perceptual experience sometimes even makes evident the resolution. My accounts of these cases avoid ascribing pervasive illusion to audition, and each reveals something significant about the mechanisms involved in auditory perception. Interference effects demonstrate that one's present perspective need not reveal all the sounds in one's surroundings—information about sounds might be lost or obscured since local waves subject to cancellation stimulate audition. Transmission through barriers, reflection, and echo experiences teach not to multiply sounds beyond sources. Doppler effects illustrate auditory constancy for pitch perception. Nonetheless, I have claimed that, where possible, the methodological model should be the following: Unless good reasons suggest we are subject to a given illusion in audition, avoid ascribing that illusion.

The event view challenges the simple visuocentric understanding according to which perception reveals just objects and their attributes. Though my account makes the place of sounds in the world unmysterious, it might be thought to add mystery to how we perceive them. Since sounds are neither qualities nor object-like particulars, the event view of sounds entails that more variety exists among the immediate objects of perception than many modern views acknowledge. If sounds are the immediate objects of auditory awareness, and sounds are events, then audition involves, in the innocuous sense discussed earlier, unmediated awareness of events. That is, awareness of a sound is awareness of an event that need not be mediated by prior awareness of some object and its attributes, states, or changes. In fact, any auditory awareness of the activities of ordinary objects is mediated by awareness of sounds. Events enter the unmediated content of auditory perception according to this account. Some might object on these grounds alone. I, however, am encouraged. Empirical evidence and philosophical arguments, such as those discussed in Chapter 5, increasingly suggest that happenings, and even causings, are far more fundamental to perceptual experience than traditional views allow. Such happenings as flashes and movements might yet find their way into accounts of immediate experience. An understanding of the contents of perception that recognizes suitable variety among the basic constituents of one's perceptual experience is poised to explain how perceivers like us satisfy the rich set of needs imposed by a complex and changing environment.

I stated at the outset that my goal for this book was not to exhaustively reconcile audition with a comprehensive theory of perception. This work's concern has been to provide the background and to inform the broader discussion by treating sounds and audition as topics worthy of interest in their own right. I am convinced that sounds are worthy subjects of our attention, and that we should devote more of it to them. Such attention promises philosophical and even aesthetic rewards (see, for instance, the ambitiously optimistic Schafer 1977). The enterprise is poised to impact the wider discussion I have had in view all along. Nevertheless, my approach here ultimately reaches limits. Looking forward, and in conclusion, I want to suggest that these limits stem from a more insidious manifestation of visuocentrism to which this approach falls prey. Let me begin with a final puzzle about audition.

11.1 A Puzzle about Audition

A substantial puzzle remains. It is clear that, in the relatively innocuous sense with which I have been working, sounds are the immediate objects of auditory experience—whatever else you hear, such as cars or crashes, you hear it in virtue of hearing a sound. Auditory experience is, nonetheless, *object-* and *event-*involving. You learn on the basis of auditory experience that the glass has broken, that there is a bell in the room, or that the train is passing. One plausible view about how you learn this is that you hear the train, the bell, or the breaking of the glass. The experience seems to you to be an experience of a train, a bell, or a glass breaking. It is no accident that we speak about and classify sounds in just these terms. So, you hear a sound, and by or in hearing that sound, you hear the everyday object or event that is its source. This awareness may feel less 'direct' than your awareness of the sound or the awareness of the apple you enjoy on the basis of seeing its color and shape, but there is a sense in which it seems that one enjoys auditory awareness of a train, a bell, or a glass breaking in virtue of hearing its sound.

The puzzle is this: How could auditory experience, whose proper objects are sounds, which are distinct from ordinary objects and events, furnish perceptual awareness of things like trains, bells, and breakings?

The puzzle raises two closely related questions about the content of auditory perception. The first is, 'How mediated is one's awareness of ordinary objects and events in audition?' That is, how remote is the sense in which one grasps such particulars as bells and whistles if that awareness is mediated by sounds? The second is, 'How rich is the content of auditory experience?' Simply, do those contents encompass bells and whistles, or merely sounds?

Defending an answer to these questions requires a story both about the contents of such awareness and about what grounds this particular form of mediated perceptual awareness. In short, we must explain how hearing could inform us perceptually about ordinary objects or events, which traditionally have not been considered among the objects of audition proper.

It should by now be clear that what we might think of as 'situating perception' is not solely the business of vision (or tactile-kinaesthetic awareness). Audition, too, serves to situate a subject with respect to the environment by furnishing a perspective on objects and events that reveals surprisingly vivid detail. But audition not only is thoroughly spatial, it also incorporates distinctively detailed temporal characteristics. Audition thereby grounds one in space and time and conveys qualitative information that complements visual awareness. It appears, therefore, to afford access to a complex world of items and occurrences.

We are concerned more often than not with things and happenings in the environment that give rise to sounds (musical listening and musical experience often are exceptions), and hearing is a rich source of such information. But how much of this information is gleaned by means that are, properly speaking, perceptual, and how much results from mere inference? I suspect that frequently we auditorily perceive ordinary objects and events by perceiving their sounds, and that we do not merely infer their presence based on past experience with similar things and their sounds. Suppose you know it is only a clever recording being played on a stereo next door and not a vacuum running. Though you are not inclined to infer or believe in the presence of a running vacuum, it still can seem to you auditorily as if a vacuum is running. There is a difference between thinking or believing that something is present and experiencing it as present. To remove awareness of the sources of sounds from auditory experience is to fail to capture perhaps the defining characteristic of that experience. To say that an auditory experience is as of a sound with such-and-such qualities leaves out a critical feature of how that sound strikes us in auditory experience—that is, as the sound *of* a particular sort of thing or happening going on in our environment.

One might argue that this apparent auditory perceptual awareness of ordinary objects and events is a mere illusion, and that the sound in fact mediates consciousness of objects and events only modulo some inferential or otherwise cognitive connection. But the phenomenology of audition seems for all the world to furnish experiential awareness of things and happenings beyond sounds. We reflexively act to orient toward or to avoid the source of a sound; research has even shown that outfielders instinctively run either forward or back to field a fly ball after hearing the bat crack while visual information about the ball's trajectory still is ambiguous (Adair 2001, 2002). Nonetheless, perhaps the puzzle depends on missing the crucial cognitive step. If you are auditorily aware only of sounds and their qualities, and if any consciousness of ordinary objects and events is mediated by some further non-perceptual cognitive states, then the puzzle dissolves apart from the question how it could strikingly seem that you are aware of commonplace objects and events in audition. Since the seeming requires explanation, even then a version of the puzzle persists.

If, however, the content of audition is very rich and audition represents, for instance, something like *train oncoming*, or *glass breaking*, in a way that involves exclusively perceptual states, then the puzzle is at its most pressing.

If the truth is somewhere in between, and audition furnishes awareness of more general categories like *source* or *object* or *event*, then the puzzle still arises. How could an ordinary material object or event be among the objects of auditory perceptual experience when sounds are in the first case the things we hear? How, in general, is it possible for awareness as of a sound to furnish awareness as of something extra-acoustic? How, that is, could audition ever represent the presence of something that is not a sound?

11.2 The Composite Snapshot Conception of Perceptual Experience

I want to suggest that the puzzle just described ultimately has its source in a widely accepted conception of the role of the senses in perception and perceptual experience that is underwritten by visuocentric thinking. Some explanation is needed.

It is fair to say that the traditional empiricist conception of overall perceptual experience is what we might call the 'composite snapshot conception' of perceptual experience, with an emphasis on 'composite'.[1] The composite snapshot conception is that perceptual experience is comprised of a set of discrete, modality-specific components superimposed (in the sense that each remains evident) to create one's total perceptual experience at a time. One's total perceptual experience is comprised of and exhausted by visual, auditory, olfactory, tactile, and gustatory experiences, each with its own distinctive modality-specific character. According to this way of understanding perceptual experience, vision might have a certain content characterized by colors and shapes, and perhaps 'visual objects'; audition's content comprises sounds, their pitches, and loudness; smell has a content characterized by odors and olfactory qualities; touch reveals textures, degrees of warmth, and pressure; and so on for each of the perceptual modalities. Compare David Lewis's characterization of the 'color mosaic' conception of visual experience:

> Those in the traditions of British empiricism and introspectionist psychology hold that the content of visual experience is a sensuously given mosaic of color spots, together with a mass of interpretive judgments injected by the subject. (Lewis 1966: 357)

Similar mosaics could be described for each sensory modality according to the composite snapshot conception of perceptual

[1] Alva Noë's (2004) discussion of what he calls the 'snapshot conception' of visual experience is unconnected with, though it served as inspiration for, this characterization of perceptual experience.

experience. Each modality, according to this traditional empiricist picture, delivers a discrete snapshot of the world from its unique perspective that is recognizably distinct from each of the others. Vision could not share elements of audition's snapshots, and vice versa. The aggregate of these snapshots, a sort of composite snapshot, constitutes and exhausts the content of one's total perceptual experience.

This traditional picture does not necessarily rule out that there could be 'common sensibles' accessible to more than one modality. For example, shape properties might be experienced through vision and through touch. What the traditional empiricist conception assumes is that these experiences are modality-specific and distinctive. There is, therefore, a distinctively visual way of experiencing shape that differs from the tactile experience of shape (see, for instance, Chalmers 2004). This animates the long history of resistance to answering affirmatively Molyneux's question whether a subject without relevant background experience could visually identify a cube formerly only felt. It also drives Bertrand Russell's extreme claim that visual, tactile, and auditory space are distinct from each other, and from the space of science:

> To begin with, space as we see it is not the same as space as we get it by the sense of touch; it is only by experience in infancy that we learn how to touch things we see, or how to get a sight of things which we feel touching us. But the space of science is neutral as between touch and sight; thus it cannot be either the space of touch or the space of sight. (Russell 1912: 29)

The traditional conception stems from thinking of the senses as distinct systems or channels of awareness. Each is understood to involve separate processes and to work in isolation from the others, at least until some relatively late stage. In addition, each modality delivers an experience with a distinctive qualitative character that could not be replicated by another modality. Each such modality delivers the experiential ingredients which, taken collectively, exhaust one's total perceptual experience.

The lesson of this chapter is that this traditional story is false in crucial respects and incomplete in others. I want to suggest that an important class of perceptual effects that has gone relatively unrecognized or unappreciated by philosophers gives us good reason to think that the composite snapshot conception of experience is incorrect. But the illusions that I shall discuss do not have just negative implications. They provide ingredients for the beginning of a solution to the puzzle about audition that I described earlier. More importantly, however, this result is not of mere parochial interest. Such illusions illuminate perhaps the most significant facet of perception—its capacity to furnish awareness as of a world of things and happenings distinct from oneself—and teach us what we could not otherwise have learned with attention restricted to vision, or to any other individual modality on its own.[2]

Thus we at last encounter the limits of the approach I have deployed throughout most of this book. Though considering audition and sounds for their own sake, as topics of interest in their own right, informs perceptual theorizing and guides us toward confronting the kinds of concerns I raise in this chapter, the risk is an auditory analog of visuocentrism. The modalities of awareness, however, cannot fully be understood individually in isolation from each other. Perception and perceptual experience are very much the results of integrating, weighing, comparing, and extracting significant information from the senses considered collectively. Perceiving is not merely a matter of assembling discrete snapshots from each modality-specific perspective. Theorizing about perception is at its most promising, therefore, when it comprehends the interactions and relationships among modalities.

[2] Trout (2001) is a notable exception that recognizes, confronts, and draws lessons unfriendly to individualist views of the mental from cross-modal effects in speech perception. Trout's conclusions, in one regard, are very much in sympathy with my arguments in this chapter. That is, what makes cross-modal effects intelligible is appeal to environmental particulars that stimulate experience in multiple sense modalities.

11.3 Cross-Modal Illusions

The perceptual effects I have in mind are ones in which what is sensed through one modality affects what is experienced in another. One familiar example, the ventriloquist illusion, has been well studied since the nineteenth century. Seeing the movements of a puppet's mouth affects where one hears the sound of a voice. But the effect is not limited to the perception of speech. Work in the second half of the twentieth century, since Howard and Templeton (1966), in particular, has confirmed various ways in which the visual location of a stimulus affects perceived auditory location. The effect is neither cognitive nor inferential, but results from cross-modal perceptual interactions (see Bertelson 1999; Vroomen et al. 2001).

Cross-modal connections similarly are revealed in several other surprising varieties of illusion. Visual capture demonstrates that sight alters, for example, the tactile perception of object size and one's proprioceptive impression of bodily orientation: seeing a hand that is not one's own results in feeling one's hand to be where it is not (Hay et al. 1965; Pick et al. 1969), and seeing a larger object affects felt object size (Rock and Victor 1964).

In the fascinating McGurk effect, subjects shown video of a speaker articulating the velar (pronounced with the back of the tongue on the soft palate)/ga/sound while presented with audio of the bilabial (pronounced with the lips)/ba/sound experience an alveolar (pronounced with the tip of the tongue behind the teeth)/da/sound (McGurk and MacDonald 1976). The McGurk effect takes place when conflicting auditory and visual information about speech is reconciled into a kind of 'average' or parsimonious percept.

Each of the preceding effects, however, could be explained in terms of vision's dominance over some other modality. In the cases I have mentioned, when sight and either audition or touch conflict, the resolution favors vision. It might be that because vision is so evolved as a 'survival sense' that whenever stimuli across modalities are incongruous, vision wins. Perhaps visuocentrism is vindicated by vision's dominance in perception over the other modalities?

Not so. Ladan Shams and her colleagues recently have discovered a class of illusions in which audition affects vision (Shams et al. 2000, 2002). In the *sound-induced flash illusion*, subjects presented with a single flash and double beep have the same visual experience as when presented with a double flash accompanied by a double beep. They see two flashes instead of one when they hear two beeps—the double auditory beep affects visual experience.

> A single flash accompanied by multiple beeps is perceived as multiple flashes. This phenomenon clearly demonstrates that sound can alter the visual percept qualitatively even when there is no ambiguity in the visual stimulus. (Shams et al. 2002: 152)

Three features of this result are significant. First, it is neither cognitive, nor inferential, nor based on some strategy adopted to respond to ambiguous or conflicting experiences. Shams et al. (2002) maintain that audition influences the phenomenology of vision as a result of cross-modal *perceptual* interactions where we otherwise might expect vision and audition each to be autonomous and univocal. Second, these and many other cross-modal effects are pre-attentional. 'Cross-modal interaction reorganizes the auditory-visual spatial scene on which selective attention later operates' (Bertelson and de Gelder 2004: 165). Finally, no semantic contribution from familiar bimodal contexts is necessary to generate the effect. It appears to be a perceptual effect grounded at a relatively low level. The effect does not result from learning for a particular context, and does not require specific bimodal experience. It is an audition-induced phenomenological change in the character of visual experience that persists through shifts in setting and stimulus characteristics.

> We present the first cross-modal modification of visual perception which involves a phenomenological change in the quality—as opposed to a small, gradual, or quantitative change—of the percept of a non-ambiguous visual stimulus. We report a visual illusion which is induced by sound:

when a single flash of light is accompanied by multiple auditory beeps, the single flash is perceived as multiple flashes. We present two experiments as well as several observations which establish that this alteration of the visual percept is due to cross-modal perceptual interactions as opposed to cognitive, attentional, or other origins. (Shams et al. 2002: 147)

11.4 Explaining Cross-Modal Illusions

The cross-modal illusions, I contend, are surprising. What makes them surprising is our allegiance to a traditional view about the role of the senses as discrete modes of perception and perceptual experience. What, then, are the consequences of cross-modal illusions for philosophical theorizing about perception, perceptual content, and perceptual experience?

Cases in which one perceptual modality affects experience in another modality are familiar from the study of synesthesia (see, for instance, Baron-Cohen and Harrison 1997; Cytowic 1998, 2002; Harrison 2001). Some people systematically and persistently experience colors when hearing sounds, experience shapes as a result of taste, or experience sounds in terms of colors, shapes, or flavors. But synesthesia is rare. (LSD-induced synesthesia may be systematic but rarely persists apart from flashbacks.) Recent estimates that put rates at roughly 1 in 22 individuals are substantially higher than previous estimates at around 1 in 2000 (see, respectively, Simner et al. 2006 and Baron-Cohen et al. 1996). Even these rare cases, however, are not ones in which subjects possess a special perceptual *capacity*, such as the ability to see into the ultraviolet range or to detect the smell of cancer. Synesthesia is something like a quirk in processing. Synesthetes do not literally perceive the color of a sound or the shape of a taste, since sounds and tastes lack those colors and shapes. Synesthetes do not literally perceive through audition the color of a sounding object or through gustation the shape of tasted food. The experience always involves an illusion.

The cross-modal illusions I have been considering are common among ordinary perceivers and do not appear to be the results of simple quirks of processing. Synesthesia differs from ordinary cross-modal illusions in that, in synesthesia but not in ordinary cross-modal interactions, information from one sense modality *accidentally* impacts experience in another. Cross-modal illusions, unlike synesthetic ones, are intelligible responses to unusual or extraordinary situations. The responses are intelligible because they result from mechanisms that are perceptually effective, and thus useful, under overwhelmingly prevalent environmental conditions. The world does not frequently contain sounds at locations that differ from the visible locations of their sources. The size of a sphere one holds usually corresponds to the size of the sphere one sees oneself as holding. Acoustic events tend to match in number the salient visible events that are their potential sources. Cross-modal interactions appear to stem from perceptual organizing principles that in general are adaptive and advantageous. Though they sometimes lead to illusions, they commonly get things right and aid to integrate sensory information.

Since these cross-modal effects are not just systematic and persistent, but also are intelligible as perceptual strategies, their explanation demands a form different from that of synesthetic illusions. To explain the influence of one modality upon what is experienced in another modality, in a way that captures the *environmental* significance of correlations across multiple modalities and, thus, their *adaptive* significance, requires appeal to some common factor that makes principles for grouping and organizing stimuli across the modalities intelligible. Notice that such intelligibility and, thus, environmental and adaptive significance seem absent from synesthetic groupings.

These considerations are reflected in what have been called *unity assumptions* for (non-synesthetic) cross-modal interactions (see Welch and Warren 1980). For example, when an incongruence (spatial or temporal) between stimuli from different modalities is relatively limited and when concordance surpasses some threshold, a common environmental source likely accounts for both stimuli.

The perceptual system's response under these conditions exhibits cross-modal biases, recalibrations, or illusions.

Unity assumptions and the cross-modal illusions they explain illustrate that visual and auditory stimuli are treated as evidence of a single environmentally significant entity or event and that a perceptual 'unit' is formed according to principles analogous to those involved in Gestalt formation from vision and from audition. The principles according to which perceptual units are formed in the cross-modal cases, however, are not limited to a single modality. Rather, they deal with the integration of information from different sensory systems. In order to avoid characterizing cross-modal illusions as accidental quirks of processing, these principles must be understood to invoke assumptions about a common environmental object or event that gives rise to sensory stimulation in multiple modalities, and about how it does so (see, for instance, de Gelder and Bertelson 2003, Wallace et al. 2004). Welch and Warren (1980: 638) argued early in the study of cross-modal interactions that to explain such effects requires 'the supposition that intersensory bias is a result of an attempt by the perceptual system to maintain a perceptual experience consonant with a unitary event'. The important point is that these unity assumptions are not specific only to a particular modality; instead, they amount to either modality-independent or multimodal assumptions concerning environment particulars that stimulate perceptual experience. They are, in effect, modality-independent assumptions about the sources of sensory stimulation. It is precisely because these grouping principles capture genuine regularities in the world of objects and events that awareness across different modalities constitutes genuine perceptual awareness of objects and events in the world.

A gap remains between influences across the modalities at the subperceptual level and the failure of the composite snapshot conception at the level of conscious perceptual awareness. Subperceptual auditory processing might result in illusory visual experiences without this showing anything about either the content (its nature, constituents, or richness) of the overall perceptual experience or the appropriateness of the composite conception of experience. What is needed is a bridge between claims about the influence of one modality upon what is experienced in another and claims about the respective contents of each modality. I believe such a connection exists.

The grouping and binding principles I have mentioned appear systematically to affect or to determine content for each modality. For example, consider the principle that slightly out-of-sync visual and auditory stimuli close enough in time probably originate from a common source. Along with general deference to audition on the temporal dimension (it has much better resolution than vision on this dimension), this might result in a visual experience that comports with the auditory stimulus, even though that visual experience differs from what it would have been in absence of the auditory stimulus. In the bimodal case, the visual and auditory experiences ultimately end up the way they do because there is a level at which perception involves the presumption that, in general, such visual and auditory stimulation very likely share a common environmental cause—a common source, object, or event. Explaining the effect any other way fails to capture why it is useful for the perceptual system to try to reconcile divergent stimuli. The question is whether assumptions concerning the unitary nature of an environmental stimulus to vision and audition merely causally determine perceptual contents that remain distinct and discrete in each modality or whether they reveal that perceptual content includes unitary constituents shared among perceptual modalities.

The perceptual system deploys principles designed to track, in a causally and counterfactually dependent way, the kinds of ordinary objects and events that lead to auditory and visual stimulation. But this assumes perception utilizes modality-independent or multimodal characterizations of such objects and events. Describing these operations involves attributing to perception some traction on ordinary objects and events in a sense that goes beyond the modality-specific notions of 'visual object' or 'auditory event' deployed within a given modality. The idea is that experience is shaped by multimodal organizing principles, and that such principles track objects and events in a multimodal or modality-independent way. Deploying unity assumptions exercises a kind of perceptual grasp upon items in the environment that are available to multiple modalities, so perception itself involves a dimension of multimodal or modality-independent content that cannot be characterized in purely auditory or purely visual terms. Given that visual experience is affected and constrained by audition, and that auditory experience is affected and constrained by vision, it is therefore plausible, in addition, to ascribe a dimension of content characterized in modality-independent or multimodal terms even to vision and audition themselves. The very same amodal content might be shared by vision and audition. Visual flashes are visible as flashes of events in the environment; audible blips are audible as blips from environmental sources.

Understanding cross-modal interactions thus grounds an argument for the claim that there are consciously accessible perceptual contents shared across modalities. Consider the accuracy or correctness conditions for a given cross-modal experience, for instance, the experience of a flash-beep pairing. Arguably, one's perceptual experience is not veridical in a world in which the visual flash and the auditory beep have entirely separate and unconnected environmental sources. There must be a single thing that is both seen and heard—a single environmental event connected with both the visible and auditory experiences—to capture what the experience seems to reveal. Furthermore, this commonality is evident even at the level of phenomenology. To illustrate the point, consider a cross-modal experience that involves *intermodal* binding. Just as you see a certain individual to be both red and square because of *intramodal* binding, you may perceive something to be both red and smooth, or red and solid, because of intermodal binding. The same object or event seems to have both visible and tactile (or visible and audible) features. For this sort of experience to be accurate, there must exist a single thing that is both red and smooth, or red and solid. A single thing must bear both visible and tactile features to account for identification across modalities. A case in which there exists one thing with the visible feature and another thing with the tactile feature fails to capture the veridicality conditions for the experience. It therefore fails as a characterization of the phenomenology of the experience. Thus, there must be an element of phenomenologically accessible perceptual content shared across modalities. The cube one holds is experienced as the very same particular one sees. The blip one hears is experienced to stem from the same event as the flash one sees. Perceptual experience therefore has a dimension of content that cannot be captured by a composite of *modality-specific*—proper or unimodal—snapshot-like contents.

It seems fair, then, to suppose that the object- or event-involving character of a given modality stems from underlying multimodal principles and content with potential for sharing across modalities. Even in the case of vision, such content cannot be explained by purely visual principles; to capture it requires appeal to relations to audition and other modalities. Likewise, audition might involve a level of content that deals with environmental particulars and that it shares with vision.

We then have a foothold on the solution to the puzzle about audition I set out earlier. Audition has an object- or event-involving character because modality-independent or multimodal principles shape auditory experience and ground a level of content that cannot be characterized in purely auditory terms. We hear sources, objects, and events, and not just sounds, pitches, and timbres, because the senses do not act as isolated systems that deliver only neat, modality-specific contents from which we learn to infer the presence of ordinary objects and events. Hearing the sounds of a bowling alley is hearing bowling balls, sliding shoes, and collisions with pins because audition shares perceptual access with vision.

I am not suggesting that the modalities of sense perception cannot be differentiated. Distinct perceptual modalities might each furnish exclusive awareness of a range of proper sensibles. Proper sensibles could include qualities, such as color, timbre, taste, and warmth, and particulars, such as odors and sounds. I do not wish to eliminate all boundaries among the senses.

What I am suggesting is that a convincing explanation of the cross-modal effects requires appeal to a dimension of perceptual content shared across the modalities. If I am right, then any snapshot that emerges within a specific modality is itself already a multimodal sculpture infused with information shaped by and gleaned from the other modalities. There is no separating off without remainder the purely auditory content or the purely visual content. Even the content of vision itself cannot exhaustively be understood entirely in isolation from the other modalities.

Not only, therefore, does the traditional empiricist conception that likens perceptual experience to a composite of discrete, modality-specific snapshots fail as a characterization of perceptual experience, but its failure reveals the second perilous flaw in the visuocentric thinking from which it stems. The tendency to take vision as an independent and representative paradigm for theorizing about perception is not just incomplete, but the approach to theorizing about perception it fosters encourages us to consider each modality as an autonomous mode of awareness and domain of inquiry. I have aimed to show that this undermines a complete understanding of perception and leaves out what is most critical for resolving long-standing philosophical debates about the possibility and grounds of perceptual access to a world of things and events. Comprehending the relationships among modalities is *helpful* in resolving puzzles about audition's object- and event-involving character. It is, however, *essential* to any satisfactory philosophical understanding of perception. The tyranny of the visual threatens to blind us to the nature, character, and scope of perceptual awareness.

References

Adair, R. K. (2001). 'The Crack-of-the-Bat: The Acoustics of the Bat Hitting the Ball.' 141st Meeting of the Acoustical Society of America, Chicago, IL.

____ (2002). *The Physics of Baseball*, 3rd edn. New York: Harper Collins.

Adelson, E. H., and Movshon, J. A. (1982). 'Phenomenal Coherence of Moving Visual Patterns.' *Nature*, 300: 523–5.

ANSI (2004). 'Acoustical Terminology.' ANSI S1.1-1994, ASA 111-1994 (R2004). New York: American National Standards Institute, Acoustical Society of America.

Aristotle (1984). *De Anima*. In J. Barnes (ed.), *The Complete Works of Aristotle: The Revised Oxford Translation*. Oxford: Oxford University Press.

Armstrong, D. M. (1961). *Perception and the Physical World*. London: Routledge and Kegan Paul.

Baron-Cohen, S., Burt, L., Smith-Laittan, F., Harrison, J., and Bolton, P. (1996). 'Synaesthesia: Prevalence and Familiality.' *Perception*, 25: 1073–9.

____ and Harrison, J. E. (1997). *Synaesthesia: Classic and Contemporary Readings*. Malden, MA: Blackwell.

Bennett, J. (1988). *Events and Their Names*. Oxford: Clarendon Press.

Berkeley, G. (1709/1975*a*). 'Essay towards a New Theory of Vision.' In M. R. Ayers (ed.), *Philosophical Works, Including the Works on Vision*. London: Dent.

____ (1713/1975*b*). 'Three Dialogues between Hylas and Philonous.' In M. R. Ayers (ed.), *Philosophical Works, Including the Works on Vision*. London: Dent.

Bertelson, P. (1999). 'Ventriloquism: A Case of Cross-modal Perceptual Grouping.' In G. Aschersleben, T. Bachmann, and J. Müsseler (eds.), *Cognitive Contributions to the Perception of Spatial and Temporal Events*, pp. 347–62. Amsterdam: Elsevier.

____ and de Gelder, B. (2004). 'The Psychology of Multimodal Perception.' In C. Spence and J. Driver (eds.), *Crossmodal Space and Crossmodal Attention*, pp. 141–77. Oxford: Oxford University Press.

Blauert, J. (1997). *Spatial Hearing: The Psychophysics of Human Sound Localization*. Cambridge, MA: MIT Press.

Block, N. (1990). 'Inverted Earth.' In J. Tomberlin (ed.), *Philosophical Perspectives*, 4. Atascadero, CA: Ridgeview.

Boghossian, P. A., and Velleman, J. D. (1989). 'Colour as a Secondary Quality.' *Mind*, 98: 81–103.

—— and —— (1991). 'Physicalist Theories of Color.' *Philosophical Review*, 100: 67–106.

Bradley, P., and Tyc, M. (2001). 'Of Colors, Kestrels, Caterpillars, and Leaves.' *Journal of Philosophy*, 98: 469–87.

Breen, N., Caine, D., Coltheart, M., Hendy, J., and Roberts, C. (2000). 'Toward an Understanding of Delusions of Misidentification: Four Case Studies.' *Mind and Language*, 15: 74–110.

Bregman, A. S. (1990). *Auditory Scene Analysis: The Perceptual Organization of Sound*. Cambridge, MA: MIT Press.

Byrne, A. (2003). 'Color and Similarity.' *Philosophy and Phenomenological Research*, 66: 641–65.

—— and Hilbert, D. (1997*a*). *Readings on Color, Volume 1: The Philosophy of Color*. Cambridge, MA: MIT Press.

—— and —— (1997*b*). 'Colors and Reflectances.' In A. Byrne and D. Hilbert (1997), *Readings on Color, Volume 1: The Philosophy of Color*. Cambridge, MA: MIT Press.

—— and —— (2003). 'Color Realism and Color Science.' *Behavioral and Brain Sciences*, 26: 3–21.

Carlile, S. (ed.) (1996). *Virtual Auditory Space: Generation and Applications*. Austin, TX: R. G. Landes.

Carr, C. (2002). 'Sounds, Signals, and Space Maps.' *Nature*, 415: 29–31.

Carroll, N. (1985). 'The Power of Movies.' *Daedalus*, 114: 79–103.

Casati, R., and Dokic, J. (1994). *La Philosopie du Son*. Nîmes: Chambon.

—— and —— (2005). 'Sounds'. In E. N. Zalta (ed.), *The Stanford Encyclopedia of Philosophy*.

—— and Varzi, A. (2006). 'Events.' In E. N. Zalta (ed.), *The Stanford Encyclopedia of Philosophy*.

Chalmers, D. J. (2004). 'The Representational Character of Experience.' In B. Leiter (ed.), *The Future for Philosophy*, pp. 153–81. Oxford: Oxford University Press.

____ (2006). 'Perception and the Fall from Eden.' In T. S. Gendler and J. Hawthorne (eds.), *Perceptual Experience*, pp. 49–125. Oxford: Clarendon Press.

Cohen, J., and Meskin, A. (2004). 'On the Epistemic Value of Photographs.' *Journal of Aesthetics and Art Criticism*, 62: 197–210.

Colburn, H. S., Shinn-Cunningham, B. S., Kidd, G., and Durlach, N. (2006). 'The Perceptual Consequences of Binaural Hearing.' *International Journal of Audiology*, 45: S34–44.

Crane, T. (1988). 'The Waterfall Illusion.' *Analysis*, 48: 142–7.

Cytowic, R. E. (1998). *The Man Who Tasted Shapes*. Cambridge, MA: MIT Press.

____ (2002). *Synesthesia: A Union of the Senses*, 2nd edn. Cambridge, MA: MIT Press.

Davidson, D. (1970). 'Events as Particulars.' In *Essays on Actions and Events*. Oxford: Clarendon Press.

de Gelder, B., and Bertelson, P. (2003). 'Multisensory Integration, Perception and Ecological Validity.' *Trends in Cognitive Sciences*, 7: 460–7.

Descartes, R. (1637/2001). 'Optics.' In P. J. Olscamp (ed.), *Discourse on Method, Optics, Geometry, and Meteorology*. Indianapolis, IN: Hackett.

Elton, C. (2006). 'Measuring Emotion at the Symphony.' *The Boston Globe*, 5 April.

Evans, G. (1980). 'Things Without the Mind—A Commentary upon Chapter Two of Strawson's *Individuals*.' In Z. van Straaten (ed.), *Philosophical Subjects: Essays Presented to P. F. Strawson*. Oxford: Clarendon Press.

Feldman, J., and Tremoulet, P. D. (2006). 'Individuation of Visual Objects over Time.' *Cognition*, 99: 131–65.

Fodor, J. (2000). 'A Science of Tuesdays.' *London Review of Books*, 22/14.

Galton, A. (1984). *The Logic of Aspect*. Oxford: Clarendon Press.

Gelfand, S. A. (1998). *Hearing: An Introduction to Psychological and Physiological Acoustics*, 3rd edn. New York: Marcel Dekker.

Gibson, J. J. (1979). *The Ecological Approach to Visual Perception*. Hillsdale, NJ: Lawrence Erlbaum.

Grice, H. P. (1961). 'The Causal Theory of Perception.' *Proceedings of the Aristotelian Society*, 35: 121–52.

Handel, S. (1995). 'Timbre Perception and Auditory Object Identification.' In B. C. Moore (ed.), *Hearing*, pp. 425–61. San Diego, CA: Academic Press.

Hardin, C. L. (1993). *Color for Philosophers*. Indianapolis, IN: Hackett.

Harman, G. (1990). 'The Intrinsic Quality of Experience.' In J. Tomberlin (ed.), *Philosophical Perspectives*, 4. Atascadero, CA: Ridgeview.

Harrison, J. (2001). *Synaesthesia: The Strangest Thing*. New York: Oxford University Press.

Hartmann, W. M., and Wittenberg, A. (1996). 'On the Externalization of Sound Images.' *Journal of the Acoustical Society of America*, 99: 3678–88.

Hay, J. C., Pick, H. L., and Ikeda, K. (1965). 'Visual Capture Produced by Prism Spectacles.' *Psychonomic Science*, 2: 215–16.

Helmholtz, H. (1877/1954). *On the Sensations of Tone*, 4th edn. New York: Dover.

—— (1925). *Treatise on Physiological Optics*, vol. III. New York: Optical Society of America.

Hilbert, D. R. (1987). *Color and Color Perception: A Study in Anthropocentric Realism*. Stanford, CA: CSLI.

Howard, I. P., and Templeton, W. B. (1966). *Human Spatial Orientation*. London: Wiley.

Jackson, F. (1986). 'What Mary Didn't Know.' *Journal of Philosophy*, 83: 291–5.

Johnston, M. (1992). 'How to Speak of the Colors.' *Philosophical Studies*, 68: 221–63.

Kepler, J. (1604/2000). *Optics*. Santa Fe, NM: Green Lion Press.

Kershaw, S. (2002). 'The Hunt for a Sniper: The Advice; Feeling That Witnesses Need a Hand, Police Offer One.' *The New York Times*, 16 October.

Kim, J. (1973). 'Causation, Nomic Subsumption, and the Concept of Event.' *Journal of Philosophy*, 70: 217–36.

Land, E. H. (1977). 'The Retinax Theory of Color Vision.' *Scientific American*, 237/6: 108–28.

Lewis, D. (1966). 'Percepts and Color Mosaics in Visual Experience.' *The Philosophical Review*, 75: 357–68.

—— (1986). 'Events.' In *Philosophical Papers, Volume II*. New York: Oxford University Press.

Locke, J. (1689/1975). *An Essay Concerning Human Understanding*. Oxford: Clarendon Press.

____ (1823). 'Elements of Natural Philosophy.' In *The Works of John Locke*, vol. III. London: Printed for Thomas Tegg.

Luce, R. D. (1993). *Sound and Hearing*. Hillsdale, NJ: Erlbaum.

McAdams, S., and Bregman, A. S. (1979). 'Hearing Musical Streams.' *Computer Music Journal*, 3/4: 26–43.

McGurk, H., and MacDonald, J. (1976). 'Hearing Lips and Seeing Voices.' *Nature*, 264: 746–8.

Maclachlan, D. L. C. (1989). *Philosophy of Perception*. Englewood Cliffs, NJ: Prentice Hall.

McLaughlin, B. P. (2003). 'The Place of Color in Nature.' In R. Mausfeld and D. Heyer (eds.), *Color Perception: Mind and the Physical World*. New York: Oxford University Press.

Malpas, R. M. P. (1965). 'The Location of Sound.' In R. J. Butler (ed.), *Analytical Philosophy*, second series, pp. 131–44. Oxford: Basil Blackwell.

Marr, D. (1982). *Vision*. San Francisco, CA: W. H. Freeman.

Michotte, A. (1963). *The Perception of Causality*. New York: Basic Books.

Mills, A. W. (1972). 'Auditory Localization.' In J. Tobias (ed.), *Foundations of Modern Auditory Theory*, vol. II, pp. 303–48. New York: Academic Press.

Newton, I. (1704/1979). *Opticks*. New York: Dover

Noë, A. (2002). 'Is the Visual World a Grand Illusion?' *Journal of Consciousness Studies*, 9/5: 1–12.

____ (2003). 'Causation and Perception: The Puzzle Unraveled.' *Analysis*, 63/2: 93–100.

____ (2004). *Action in Perception*. Cambridge, MA: MIT Press.

Nudds, M. (2001). 'Experiencing the Production of Sounds.' *European Journal of Philosophy*, 9: 210–29.

O'Callaghan, C. (2002). *Sounds*. PhD thesis, Princeton University.

____ (2007*a*). 'Echoes.' *The Monist*, 90/3. (forthcoming)

____ (2007*b*). 'Perceiving the Locations of Sounds.' *European Review of Philosophy*, 7. (forthcoming)

Palmer, A. R. (1995). 'Neural Signal Processing.' In B. C. J. Moore (ed.), *Hearing*, 2nd edn, pp. 75–121. San Diego, CA: Academic Press.

Palmer, S. E. (1999). *Vision Science: Photons to Phenomenology*. Cambridge, MA: MIT Press.

Pasnau, R. (1999). 'What is Sound?' *Philosophical Quarterly*, 49: 309–24.
—— (2000). 'Sensible Qualities: The Case of Sound.' *Journal of the History of Philosophy*, 38: 27–40.
Pick, H. L., Warren, D. H., and Hay, J. C. (1969). 'Sensory Conflict in Judgments of Spatial Direction.' *Perception and Psychophysics*, 6: 203–5.
Reé, J. (1999). *I See a Voice*. London: Harper Collins.
Rock, I. (1983). *The Logic of Perception*. Cambridge, MA: MIT Press.
—— and Victor, J. (1964). 'Vision and Touch: An Experimentally Created Conflict between the Two Senses.' *Science*, 143/3606: 594–6.
Russell, B. (1912). *The Problems of Philosophy*. London: Oxford University Press.
Schafer, R. M. (1977). *The Tuning of the World*. New York: Alfred A. Knopf.
Scholl, B., and Nakayama, K. (2004). 'Illusory Causal Crescents: Misperceived Spatial Relations due to Perceived Causality.' *Perception*, 33: 455–69.
—— and Pylyshyn, Z. (1999). 'Tracking Multiple Items through Occlusion: Clues to Visual Objecthood.' *Cognitive Psychology*, 38: 259–90.
—— and Tremoulet, P. (2000). 'Perceptual Causality and Animacy.' *Trends in Cognitive Sciences*, 4: 299–309.
Schouten, J. F. (1940). 'The Residue, a New Concept in Subjective Sound Analysis.' *Proceedings of the Koninklijke Nederlandse Akadademie*, 43: 356–65.
Sekuler, R., Sekuler, A. B., and Lau, R. (1997). 'Visual Influences on Auditory Plunk and Bow Judgments.' *Perception and Psychophysics*, 54: 406–16.
Shams, L., Kamitani, Y., and Shimojo, S. (2000). 'What You See Is What You Hear.' *Nature*, 408: 788.
—— —— and —— (2002). 'Visual Illusion Induced by Sound.' *Cognitive Brain Research*, 14: 147–52.
Shinn-Cunningham, B. (2001*a*). 'Creating Three Dimensions in Virtual Auditory Displays.' In M. Smith, G. Salvendy, D. Harris, and R. Koubek (eds.), *Usability Evaluation and Interface Design: Cognitive Engineering, Intelligent Agents and Virtual Reality*, pp. 604–8. Hillsdale, NJ: Lawrence Erlbaum.

——(2001*b*). 'Localizing Sound in Rooms.' In *Acoustic Rendering for Virtual Environments*, pp. 17–22. ACM SIGGRAPH.

——(2004). 'The Perceptual Consequences of Creating a Realistic, Reverberant 3-D Audio Display.' *Proceedings of the International Congress on Acoustics*, Kyoto, Japan.

——(2005). 'Influences of Spatial Cues on Grouping and Understanding Sound.' In *Proceedings of the Forum Acusticum*.

Shoemaker, S. (1982). 'The Inverted Spectrum.' *Journal of Philosophy*, 79: 357–81.

Siegel, S. (2005). 'The Phenomenology of Efficacy.' *Philosophical Topics*, 33: 265–84.

——(2006*a*). 'Which Properties Are Represented in Perception?' In T. S. Gendler and J. Hawthorne (eds.), *Perceptual Experience*, pp. 481–503. Oxford: Clarendon Press.

——(2006*b*). 'Subject and Object in the Contents of Visual Experience.' *Philosophical Review*, 115: 355–88.

——(2008). 'The Visual Experience of Causation.' *Philosophical Quarterly*.

Simner, J., Mulvenna, C., Sagiv, N., Tsakanikos, E., Witherby, S. A., Fraser, C., Scott, K., and Ward, J. (2006). 'Synaesthesia: The Prevalence of Atypical Cross-modal Experiences.' *Perception*, 35: 1024–33.

Snowdon, P. (1992). 'How to Interpret "direct perception".' In T. Crane (ed.), *The Contents of Experience*, pp. 48–78. Cambridge: Cambridge University Press.

Strawson, P. F. (1959). *Individuals*. New York: Routledge.

——(1974). 'Causation in Perception.' In *Freedom and Resentment and Other Essays*, pp. 66–84. London: Methuen.

Terhardt, E. (1974). 'Pitch, Consonance and Harmony.' *Journal of the Acoustical Society of America*, 55: 1061–9.

——(1979). 'Calculating Virtual Pitch.' *Hearing Research*, 1: 155–82.

Thomson, J. J. (1983). 'Parthood and Identity across Time.' *The Journal of Philosophy*, 80: 201–20.

Trout, J. D. (2001). 'Metaphysics, Method, and the Mouth: Philosophical Lessons of Speech Perception.' *Philosophical Psychology*, 14: 261–91.

Tye, M. (2000). *Consciousness, Color, and Content*. Cambridge, MA: MIT Press.

Vitello, P. (2006). 'A Ring Tone Meant to Fall on Deaf Ears.' *The New York Times*, 12 June.

Vroomen, J., Bertelson, P., and de Gelder, B. (2001). 'Auditory-Visual Spatial Interactions: Automatic versus Intentional Components.' In B. de Gelder, E. de Haan, and C. Heywood (eds.), *Out of Mind*, pp. 140–50. Oxford: Oxford University Press.

Wallace, M. T., Roberson, G. E., Hairston, W. D., Stein, B. E., Vaughan, J. W., and Schirillo, J. A. (2004). 'Unifying Multisensory Signals across Time and Space. *Experimental Brain Research*, 158: 252–8.

Walton, K. (1984). 'Transparent Pictures: On the Nature of Photographic Realism.' *Critical Inquiry*, 11: 246–76.

Watanabe, K., and Shimojo, S. (2000). 'Postcoincidence Trajectory Duration Affects Motion Perception.' *Perception and Psychophysics*, 9: 557–64.

Weiskrantz, L. (1986). *Blindsight: A Case Study and Implications*. Oxford: Clarendon Press.

Welch, R. B., and Warren, D. H. (1980). 'Immediate Perceptual Response to Intersensory Discrepancy.' *Psychological Bulletin*, 88: 638–67.

Westfall, R. S. (1980). *Never at Rest: A Biography of Isaac Newton*. Cambridge: Cambridge University Press.

Woolsey, C. N. (1960). 'Organization of the Cortical Auditory System: A Review and Synthesis.' In G. Rasmussen and W. Windle (eds.), *Neural Mechanisms of the Auditory Vestibular System*, pp. 165–80. Springfield, IL: Thomas.

Xu, F. (1999). 'Object Individuation and Object Identity in Infancy: The Role of Spatiotemporal Information, Object Property Information, and Language.' *Acta Psychologia*, 102: 113–36.

Zahorik, P., and Wightman, F. (2001). 'Loudness Constancy with Varying Sound Source Distance.' *Nature Neuroscience*, 4: 78–83.

Index